Forest and Plant Bioproducts Division 2017

Core Programming Area at the 2017 AIChE Annual Meeting

Minneapolis, Minnesota, USA
29 October – 3 November 2017

ISBN: 978-1-5108-5804-6

Printed from e-media with permission by:

Curran Associates, Inc.
57 Morehouse Lane
Red Hook, NY 12571

Printed by Curran Associates, Inc. (2018)

For permission requests, please contact AIChE
at the address below.

AIChE
120 Wall Street, FL 23
New York, NY 10005-4020

Phone: (800) 242-4363
Fax: (203) 775-5177

www.aiche.org

Additional copies of this publication are available from:

Curran Associates, Inc.
57 Morehouse Lane
Red Hook, NY 12571 USA
Phone: 845-758-0400
Fax: 845-758-2633
Email: curran@proceedings.com
Web: www.proceedings.com

TABLE OF CONTENTS

Some abstracts in this book refer to images that were not made available by the publisher

499972 Bioprocess Model and Economic Analysis of Microalgae Production in Flat-Panel Photobioreactors Taking into Account Geospatial Factors

Abstract:

Microalgae are a promising feedstock for production of fuels and value-added chemicals. In order to assess the economic, energetic and environmental performance of algae based system in an industrial scale biorefinery, it is important to model algae productivity accurately. In this work, a first principle based growth model of microalgae is developed for flat panel photobioreactors taking into account the transport and biological processes. The primary factors that affect microalgae growth, such as solar irradiance, temperature, specific characteristics of species, uptake rates of nutrients and carbon dioxide, are incorporated in the bioreactor process model. Year-round algae productivity potential is computed for different locations to elucidate the implications of climate variations for the locations. Based on the yearly microalgal productivity for different regions, economic analysis is conducted to quantify the microalgae production cost for the different locations. As there are significant uncertainties associated with the different parameters of the growth model and economic analyses, detailed sensitivity analyses are performed. Results show that microalgae production cost is heavily influenced by climatic factors like solar irradiance and air temperature as well as by different equipment and process design factors. Results from the dynamic model can be used to optimize and scaling up of algae systems for efficient production of fuels and chemicals for variety of algae species as well as geospatial locations.

497873 Continuous Photobioreactor Cultivation of *Nannochloropsis oculata* to Isolate Cosmetical Grade Phospholipids

Abstract:

Microalgae are a promising source of natural ingredients as their cell mass contains polysaccharides, proteins, lipids, pigments, and vitamins. Phospholipids, in particular, are of interest to the cosmetics industry for various formulations, such as emulsifiers, liposome formers, solubilizers, wetting agents, and bioactive compounds. They are currently extracted from food sources, but this practice raises sustainability questions because of the competition with food. Microalgae can serve as a source of phospholipids on a more sustainable basis. To advance the status of microalgae bioproducts for cosmetics applications, the marine microalga *Nannochloropsis oculata* was cultivated in a vertical flat panel photobioreactor (VFPPBR) in batch and continuous mode. Batch cultivation was employed to optimize the growth parameters of *N. oculata*, calculate its maximum specific growth rate (μ_{max}), and determine its phospholipid productivity. The measured μ_{max} of 0.015 hr^{-1} was then used to select the dilution rate for designing and operating a VFPPBR-based continuous cultivation that is deemed more scalable and productive from a commercialization standpoint. Cell harvesting was followed by organic solvent extraction to solubilize and recover the algal phospholipids. The experimental findings in continuous cultivation of *N. oculata*, the robustness of the strain, and its polar lipid productivity, bioreactor designs, performance of the strain in batch and continuous reactors, and green pathways for downstream processing to isolate, concentrate and fractionate algal phospholipids were analyzed and discussed.

498781 Effects of pH on Cell Growth, Lipid Production and CO_2 Demand of Microalgae *Chlorella sorokiniana*

Abstract:

Microalgae have emerged as one of the most promising alternative energy feedstocks. Some advantages include the simple cellular structure, short production cycle, high lipid content, and fast growth. However, high production costs due to high CO_2 usage and low lipid productivity have been one of the major challenges impeding the commercial production of algal biofuels, Here, cell growth and lipid contents of *Chlorella sorokiniana* DOE 1412 were tested at different pH in flask cultivation. Culture pH was manipulated by CO_2 addition. The optimal pH for DOE1412 is 5.8-6.0 when only taking into account the cell growth and lipid production. A flat panel airlift photobioreactor (PBR) was used for scale-up cultivation at five different pH levels (6.5, 7.0, 7.5, 8.0 and 8.5). Data of pH value and the CO_2 addition was collected by a data logger. Biomass productivity increased with decreasing pH. By taking into account not only the cell growth and lipid production but also CO_2 demand, the lowest value of g CO_2/g dry biomass was achieved at pH 8. Also, the fatty acid profiles and biodiesel properties in different culture pH were investigated. Cetane number (CN) of biodiesel in the pH 6.5, 7 and 7.5 treatments satisfied the US standard ASTM D6751; among them, the pH 6.5 treatment met the Europe standard EN 14214. Finally, protein content in algae biomass increased with increasing pH, while C/N ratio in cells decreased.

484990 Design of Marine Macroalgae Photobioreactor Integrated into Building with Natural Sun Illumination

Abstract:

Seagriculture, which can provide offshore grown macroalgae biomass for food, platform chemicals and biofuels production will play a significant role in bioeconomy and biorefineries. Nevertheless, seagriculture development has been hindered by the lack of laboratory photobioreactors that enable fundamental macroalgae research. Here we report on the indoor macroalgae photobioreactor (MPBR) system integrated into the LEED certified building that enables maintenance and cultivation of marine macroalgae at controlled environment under natural illumination. Depending on specific reactor location, illumination varied between reactors in 238-348 µmole photons m^{-2} s^{-1} . The total direct energy consumption of the system was 8.15-11.45kWh during the 24 hours of operationThe total operation volume of the system was 6,400 L of seawater (salinity 3.9%, pH 8.2). Initial water stock was provided from the Tel Aviv Reading marine area and was partially (1/4-1/3) exchanged every 3 months. The compensation for evaporation was done by adding deionized water to the system to maintain the salinity at 3.9%.

The MPBR operation was been demonstrated for 6 months on continuous maintenance and cultivation of three green macroalgae species *Cladophora sp.*, *Ulva compressa* and *Ulva rigida* isolated form the Eastern Mediterranean coast. The growth rates, protein, rhamnose, xylose, arabinose, glucose, galactose and glucuronic acid content of the cultivated species were quantified. *Cladophora sp.*, isolated from Rosh-ha-Nikra sampling site was cultivated for 122 days. The growth rates changed from -3.66 $\%day^{-1}$ to 9.78 $\%day^{-1}$, with an average 3.47 $\%day^{-1}$. The maximum growth was for 12 continuous days. *Cladophora sp.* isolated from Mikhmoret sampling site was cultivated for 122 days. The growth rates changed from 9.32$\%day^{-1}$ to 95.5 $\%day^{-1}$, average 36.63 $\%day^{-1}$. The maximum growth was for 12 continuous days. *Cladophora sp.* isolated from Reading sampling site was cultivated for 97 days. The growth rates were stable at 95.47 $\%day^{-1}$ during the entire period of 38 cultivation days.

Carbohydrate profiling was done for *Cladophora sp.* harvested at Mikhmoret site and cultivated at the system at the growing rate of 63.9$\%^{-1}$ during harvesting. Rhamnose content was 3.89 µg mg_{DW}^{-1}, arabinose 30.46 µg mg_{DW}^{-1}, galactose 26.59 µg mg_{DW}^{-1}, glucose 61.72 µg mg_{DW}^{-1}, xylose 17.79 µg mg_{DW}^{-1} and glucuronic acid 0.98 µg mg_{DW}^{-1}. The total carbohydrate yield was 14.10%. The protein content in the biomass was 11%.

U. compressa isolated from Rosh-ha-Nikra sampling site was cultivated for 187 days. The growth rates changed from -0.06% day^{-1} to 3.5 $\%day^{-1}$, average 1.01 $\%day^{-1}$. The maximum growth was for 9 continuous days. *U. compressa* isolated from Mikhmoret sampling site was cultivated for 172 days. The growth rates changed from 1.46 $\%day^{-1}$ to 5.51 $\%day^{-1}$, average 3.65 $\%day^{-1}$.The maximum growth was for 27 continuous days. *U.compressa* isolated from Reading sampling site was cultivated for 143 days. The growth rates were mostly negative, with the biomass loss at -82.34 $\%day^{-1}$, growth rate of only 0.83 $\%day^{-1}$ was observed 40 continuous days.

Carbohydrate and protein profiling was done for *U. compressa* harvested at Rosh-ha-Nikra site and cultivated at the system at the growing rate of 3.5% day^{-1} at the time of a harvest. Rhamnose content was 48.58 µg mg_{DW}^{-1}, arabinose 0.29 µg mg_{DW}^{-1}, galactose 9.03 µg mg_{DW}^{-1}, glucose 75.13 µg mg_{DW}^{-1}, xylose 23.83 µg mg_{DW}^{-1} and glucuronic acid 18.13 µg mg_{DW}^{-1}. The total carbohydrate yield was 17.50%. The total protein content of this biomass was 25.5%.

U. rigida isolated from Rosh-ha-Nikra sampling site was cultivated for 89 days. The growth rates changed from -3.13 $\%day^{-1}$ to 19.51 $\%day^{-1}$, average 4.61 $\%day^{-1}$. The maximum growth was for 16 continuous days. *U.* rigida isolated from Haifa sampling site was cultivated for 155 days. The growth rates changed from -0.09 $\%day^{-1}$ to 9.4 $\%day^{-1}$, average 2.01 $\%day^{-1}$. The maximum growth was for 64 continuous days.

Carbohydrate and protein profiling was done for *U. rigida* harvested at Haifa site and cultivated at the system at the growing rate of 1.4%/day at the time of a harvest. Rhamnose content was 37.29 µg mg_{DW}^{-1}, arabinose 0.46 µg mg_{DW}^{-1}, galactose 3.60 µg mg_{DW}^{-1}, glucose 54.84 µg mg_{DW}^{-1}, xylose 9.11µg mg_{DW}^{-1} and glucuronic acid 16.62 µg mg_{DW}^{-1}. The total carbohydrate yield was 12.20%. The protein content from the biomass harvested was 29%.

This work provides detailed insights on the design of practical indoors photobioreactors for isolation, selection, cultivation maintenance and analysis of macroalgae species required for biorefinery feedstock development.

488865 An Integrated Approach for Bioenergy Production from Microalgae Using Solar Energy, CO_2 and Wastewater

Abstract:

New biomass sources for alternative fuels has grown to be a subject of increasing importance as the world struggles to resolve the economic and strategic impacts of limited fossil fuel resources on environment and global climate. Microalgae are among the most promising non-food-crop-based biomass feedstocks. Microalgae are photosynthetic microorganisms with simple growing requirements (light, CO_2 and nutrients) that can produce lipids, proteins and carbohydrates in large amounts over short periods of time. These products can be processed into both bioenergy and valuable co-products. On the other hand, carbon dioxide (CO_2) is one of the most important contributors for the increase of the greenhouse effect. CO_2 concentrations are increasing in the last decades mainly due to the increase of emissions due to human activities. A promising technology is the biological capture of CO_2 using microalgae. These microorganisms can fix CO_2 using light with efficiency ten times greater than terrestrial plants.

To be economically competitive with carbon capture/sequestration methodologies, an intensive research is needed to integrate all microalgal culture benefits: flue gas and wastewater treatment and biomass production. In this research, an integrated process is introduced to produce bioenergy from microalgal biomass combined with simultaneous CO_2 capture/sequestration and bioremediation of wastewater nutrients. Also, this research investigates different processes and operations to optimize microalgal biomass production using open ponds or photobioreactors. Finally, this research focuses on the potential of producing biogas from microalgal biomass through anaerobic digestion and further upgrading of biogas into higher alkenes through catalytic processes

The main objective of this research is to investigate the growth kinetics of the appropriate microalgae species at different feed composition of CO_2 with air to optimize the growth rate of microalgae at maximum CO_2 consumption and in various wastewater media to reduce wastewater nutrients concentrations (COD, ammonia, nitrate and phosphate) to the potable clean water level. Also, this research aims to study the growth kinetics of the microalgae at other important parameters such as the effect of concentrated solar energy, pH, temperature, wastewater dilution and algae circulation rate. The anaerobic digestion of the harvested algae is investigated at different operating conditions such as liquid/solid ratio, temperature and pH. Methane concentration in the outlet stream will be optimized as function of these operating conditions.

There is a number of challenging issues in the realization of the proposed technology. One of the major challenges is to find appropriate microalgae species tolerant to high concentrations of CO_2. The flue gas of a power plant contains up to 500 times higher CO_2 content than that in the atmosphere. Therefore, different microalgae species will be studied under similar composition of CO_2 in the flue gas from a fossil fuel-based power generating station to find the proper type of microalgae adapting to higher CO_2 concentrations. Another major challenge is to find the proper technology

for microalgal biomass production. There are two main technologies; open ponds and photobioreactors. Even that the open ponds are relatively cheap and easy to clean and maintain, they have poor biomass productivity, mixing, light and CO_2 utilization and they require large area of land. On the other hand, photobioreactors have different designs such as tubular, flat plate and column. Usually, the photobioreactors produce high microalgal biomass. Therefore, this research will explore different designs of open ponds and photobioreactors to optimize the growth of desired microalgae species. Finally, integrated processes propose different challenges that need to be faced. One of the powerful tools to meet these challenges is modelling and simulation software. For example, an essential part of any integrated process is heat and energy integration. Aspen simulation packages proved their capabilities to design and simulate such integration. In this research, modelling and simulation packages will be used to optimise the overall proposed integrated process along with analysing and modelling collected data from the experimental work.

It is expected that the innovation and implementation of the integrated approach for microalgal bioenergy production CO_2 will offer a solution of providing sustainable energy demand and addressing the global commitment to reduce greenhouse gases emissions by CO_2 capture/sequestration.

491918 Rapid Wood Fractionation ≤ 80°C for Sustainable and Economic Biorefinery

Abstract:

Here we report a rapid and low temperature (≤ 80°C) wood fractionation process for producing wood pulp, cellulose and lignin nanomaterials, chemicals, and sugar/biofuels. Approximately 90% of wood lignin can be dissolved at 80°C for 20 min. Equivalent delignification using alkaline pulping can be achieved only at 170°C for 2 h. The novel process fractionated wood into two fractions: (1) a primarily carbohydrate-rich water-insoluble solid fraction that can be used for sustainable production of fibers such as dissolving pulp, lignocellulosic nanomaterials, and/or sugars through subsequent enzymatic hydrolysis, and (2) a spent acid liquor stream containing mainly dissolved lignin that can be easily precipitated as lignin nanoparticles simply by diluting the spent acid liquor. The dissolved hemicellulose sugars can be used to produce furfural after lignin precipitation. The chemical used has a low water solubility, which facilitates efficient recovery simply using commercially proven crystallization technology by cooling concentrated spent acid solution to ambient temperatures to achieve environmental sustainability.

502673 Insights into Biomass Recalcitrance

Abstract:

The utilization of biomass for biofuels and bio-derived chemicals and materials based on the biological platform is contingent on efficient separation, deconstruction and conversion. Todayâ€™s cellulosic ethanol plants, especially those based on the cellulase deconstruction, have demonstrated that the conversion of plant polysaccharides to ethanol can be accomplished on an industrial scale but current process challenges are hindering wide-spread adaption. One of the key issues is the recalcitrance of biomass and the need to utilize aggressive pretreatment technologies. Recent advances in our understanding of the plant cell wall structure and our ability to reduce biomass recalcitrance by altering these constituents by controlling genetic features of the plant and innovations in pretreatment chemistries provides innovative pathways to facilitate the biorefining of plant polysaccharides to biofuels and yield a lignin rich process stream that can be valorized to chemicals and/or materials. This seminar review our contributions in this field and highlight future opportunities.

492489 Low Temperature and High Efficiency Biomass Fuel Cell and Bio-Hydrogen Production

Abstract:

Title: Low temperature and high efficiency biomass fuel cell and bio-hydrogen production

Name: Yulin Deng

School of Chemical & Biomolecular Engineering, Georgia Institute of Technology, Atlanta, GA 30302-0620, yulin.deng@rbi.gatech.edu

A novel fuel cell which can directly use native polymeric biomasses, such as starch, cellulose, lignin, and even switchgrass and wood powders will be discussed. This fuel cell combines some features of solar cells, fuel cells, and redox flow batteries. Specifically, an oxidation/reduction ion pair, such as Fe^{2+}/Fe^{3+}, or polyoxomatelate, are used as catalyst to oxidize biomass by either absorbing solar light or heat energy. The power density of the solar-induced hybrid fuel cell powered with raw biomass could reach ~50 mW/cm^2. Unlike most cell technologies that are sensitive to impurities, the cell reported in this study is inert to most organic and inorganic contaminants present in the fuels. The fuel cell is completely noble metal free. The similar fuel cell system can also be converted to electrolysis cell for hydrogen production at low temperature using native biomass directly.

504077 Circular Economy: A Path Towards Innovation and Commercialization of Biocomposites for Sustainable Manufacturing

Abstract:

“Circular economy” is an emerging concept that targets waste minimizations through a closed loop economy system, unlike the traditional linear model (“make and dispose” type economy), and therefore supports sustainable development. The underlying principle of this emerging economy is “Nothing is waste; waste is a co-product waiting for another opportunity”. In this system, the byproducts and co-products of one industry can find commercially viable applications in another industry, with appropriate design and engineering. There is a significant potential for utilization of biobased coproducts from agro and forest industries, food and beverage industry as well as biofuel and biomass-based pyrolysis industries. Our focus is on biocomposite applications that utilize these materials as reinforcing and filler materials for plastics, including biobased and biodegradable plastics. The developed eco-friendly biocomposites are suitable for use in consumer products, light-weight auto-parts and 100% compostable packaging.

Acknowledgements: Agriculture and Agri-Food Canada and Competitive Green Technologies (CGtech) through AgriInnovation Program (Project #052882 and 051910); the Ontario Ministry of Agriculture, Food, and Rural Affairs - University of Guelph Bioeconomy Industrial uses Research Program (Project # 200425); Natural Sciences and Engineering Research Council, Canada for the Discovery Grants (Project #401111); and the Ontario Research Fund, Research Excellence Program; Round-7 (ORF-RE07) from the Ontario Ministry of Research and Innovation (MRI), currently known as the Ontario Ministry of Research, Innovation and Science (MRIS) (Project # 052665 and #052644).

503670 Digital Alchemy for Assembly Engineeringc

Abstract:

Back in the day, Stuart Churchill was a pioneer in the use of computers for modeling chemical and thermal processes. Since that time, the role of computing in chemical engineering has become indispensable. In this talk, we present new computational methods we call "digital alchemy" that allow for predictive modeling of thermodynamic assembly, and permit the design of building blocks optimized for assembly into target structures with desired properties. In this approach, standard thermodynamic ensembles are generalized to include alchemical variables and their thermodynamic conjugates, allowing for rapid, real-time computational investigation of large, complex design spaces.

503671 Improved Algebraic, Numerical, and Graphical Representations in Fluid Mechanics

Abstract:

The concept of equivalents is applied to fully developed flows through straight channels, and two applications are introduced: the equivalents that produce equal rates of flow through channels of different cross-section, and the diminishment that occurs between the regimes of laminar and turbulent flow. The validity of the eddy diffusivity and the invalidity of the mixing-length as measures of the shear stress due to turbulence have perhaps never before been demonstrated so clearly. The possible choices of primary variables and the consequences are also examined. The objective of these explorations is an improved understanding of the basic fluid mechanics by teachers, students, and industrial practitioners.

503672 The Scaling of Turbulence Near the Wall and the Churchill Turbulent Flux Correlation: Insights with Lagrangian Simulations

Abstract:

The evaluation of mean velocity and mean temperature in wall-bounded turbulent flows has been based on the "classical" theory of turbulence and the law of the wall. The viscous wall parameters, namely the friction velocity, the friction temperature, and the kinematic viscosity of the fluid have been used to scale not only the mean values of velocity, but also other mean turbulent quantities close to the wall, including the Reynolds stresses and the turbulent velocity and temperature fluctuations. Stuart Churchill, as he questioned the validity of the Reynolds analogy between momentum and heat or mass transfer in his works [1], he realized that extending the scaling of momentum to the scaling of scalar quantities in turbulence was not a sound approach. He proposed a scaling of turbulence that depends on the turbulent momentum flux and the turbulent heat flux. In his AIChE Institute Lecture (in Miami, 1998), Churchill presented a summary of his work on this topic, and closed form solutions for the mean velocity and temperature [2]. The turbulent Prandtl number, the most significant parameter for the modeling of turbulent transport not only in theory but also in modern CFD commercial software packages, was also interpreted physically based on these ideas [3]. We have used direct numerical simulation results and results from Lagrangian simulations [4] for the transport of passive scalars in turbulent channel flows to explore Churchill's ideas on turbulent flow scaling and to develop an updated theory for the scaling of heat transport in turbulent flows [5-7]. His passion to remove empiricisms and approximations from the development of predictive correlations in turbulent transport has been an inspiration. In this talk, we will revisit the Churchill scaling in view of our simulation results, and will discuss the counterintuitive finding that wall turbulence can be used to separate rather than mix substances under special circumstances [8].

1. Churchill, S. W., 1996, "A Critique of Predictive and Correlative Models for Turbulent Flow and Convection," Ind. Eng. Chem. Res., 35, pp. 3122–3140.
2. Churchill, S.W., 2000, "Progress in the Thermal Sciences: AICHE Institute Lecture," AIChE Journal, 46(9), pp. 1704-1722.
3. Churchill, S. W., 2002, "A reinterpretation of the Turbulent Prandtl Number," Ind. Eng. Chem. Res., 41, pp. 6393–6401.
4. Papavassiliou, D.V, 2002, "Turbulent transport from continuous sources at the wall of a channel," Int. J. Heat Mass Transfer, 45(17), pp. 3571-3583.
5. Le, P.M., and Papavassiliou, D.V., 2008, "On the scaling of heat transfer using thermal flux gradients for fully developed turbulent channel and Couette flows," Int. Commun. Heat Mass Transfer, 35(4), pp. 404-412.
6. Srinivasan, C., and Papavassiliou, D. V., 2011, "Prediction of Turbulent Prandtl Number in Wall Flows With Lagrangian Simulations," Ind. Eng. Chem. Res., 50(15), pp. 8881–8891.
7. Srinivasan, C., and Papavassiliou, D.V., 2013, "Heat transfer scaling for wall bounded turbulent flows," Applied Mechanics Review 65(3), Art. 031002.

8. Nguyen, Q., Srinivasan, C., and Papavassiliou, D.V., 2015, "Flow induced separation in wall turbulence", Phys Rev E, 91, Art. 033019.

503674 Flow Boiling Using a Piranha Pin Fin Heat Sink

Abstract:

Flow Boiling using a Piranha Pin Fin Heat Sink

Corey Woodcock, Xiangfei Yu, Yoav Peles[1] and Joel Plawsky

Rensselaer Polytechnic Institute; [1]University of Central Florida

The *Piranha Pin Fin* (*PPF*) microdevice is a heat exchanger engineered to dissipate extreme thermal loads using either dielectric fluids such as HFE7000 or water. Fabrication of the device is CMOS compatible and the first generations of the device have been fashioned from silicon substrates. The PPF operates on the principle that heat transfer would be more efficient if heated fluids could be removed nearly as fast as they are created. This process would keep the driving force as high as possible. Thus the heat exchanger is built as two layers; a primary substrate that dissipates the bulk of the heat and shunts the heated fluids to a booster section that continues to heat the fluid and extends the total surface area of the device. Three generations of devices were fabricated. The first two generations were single-layer systems designed to help understand the operation of the primary layer and optimize the shape and location of the pin fins. The third generation was the first, two-layer system to be fabricated and tested.

Figure 1 Representative PPF array

Correlations for the heat transfer coefficient, h_{tp}, based on the Boiling Number and the Weber Number in the PPF design was developed and validated with Gen II. The discrepancy of heat transfer coefficient between predicted and measured values was within 20%. The ratio of two-phase to single-phase pressure drop was correlated with the Boiling Number (Bo) and the Jakob Number (Ja). The discrepancy of the two-phase multiplier between prediction and experimental measurements were within 12%.

The third generation *PPF* heat sinks were the first series to provide both a primary substrate populated with *PPF*s, as well as a booster substrate populated with solid micro pin fins. The substrates are micro-aligned and fusion-bonded to provide a full three dimensional (3D) architecture with bulk silicon properties. Experimental results from the Gen III *PPF* microdevices provide a proof of concept of the true *PPF* heat sink concept with fluid venting into multiple substrates. At low flowrates (G = 670 [kg/s/m^2]) the Gen III *PPF* heat sink has been shown to dissipate heat loads approaching 1.1 [kW/cm^2] at an exit quality of 95%, while at higher flowrates (G = 3,300 [kg/s/m^2]) the thermal limit of the heat sink remains unknown due to exceeding the available power supply at >1.4 [kW/cm^2].

503673 Transport Problems in the Spirit of Stuart Churchill for Teaching and Research at the University of Michigan

Abstract:

Stuart Churchillâ€™s solutions to important heat transport problems have greatly influence both pedagogy and research. The correlation of Churchill and Chu for the natural convection heat transfer to or from a vertical plate, over an astonishing range of Rayleigh numbers is based partly on a similarity solution of a type common in heat, mass, and momentum transfer for problems in chemical engineering. One such problem is asphaltene or wax transport and deposition onto the wall of an oil pipeline, which can be solved using similar methods, as will be discussed. In addition, such problems, and the wealth of their applications make for excellent motivations and examples in the teaching of graduate or undergraduate transport, as will be discussed in the talk.

509054 Solve this! Fundamental Approach to Problem Solving in Industrial Processes I (Invited Talks)

Abstract:

Solving complex plant problems is an everyday exercise that affects all projects and budgets, and it is very rarely introduced to students at any level. In fact, by the time young engineers are few years into working in industry, most have lost the engineering skills that they developed in school. We hope to encourage future engineers to retain those skills and learn the importance of plant problem-solving.

504671 Rapid and near-complete dissolution of wood lignin at â‰¤ 80Â°C using a recyclable acid hydrotrope for sustainable production of high-value building blocks

Abstract:

Here we report the discovery of the hydrotropic properties of a recyclable aromatic acid, *p*-toluenesulfonic acid (*p*-TsOH), for low-cost and efficient fractionation of wood through rapid and near-complete dissolution of lignin. Approximately 90% of poplar wood (NE222) lignin can be dissolved at 80°C for 20 min. Equivalent delignification using known hydrotropes such as aromatic salts can be achieved only at 150°C or higher for over 10 h, or at 170°C for 2 h with alkaline pulping. *p*-TsOH fractionated wood into two fractions: (1) a primarily carbohydrate-rich water-insoluble solid fraction that can be used for sustainable production of high-value building blocks, such as dissolving pulp fibers and lignocellulosic nanomaterials, and/or sugars through subsequent enzymatic hydrolysis, and (2) a spent acid liquor stream containing mainly dissolved lignin that can be easily precipitated as lignin nanoparticles simply by diluting the spent acid liquor to below the minimal hydrotrope concentration. *p*-TsOH has a low water solubility, which facilitates efficient recovery simply using commercially proven crystallization technology by cooling concentrated spent acid solution to ambient temperatures to achieve environmental sustainability.

490632 Tuning the Physicochemical Properties of Biochar Derived from Ashe Juniper By Vacuum Pressure and Temperature

Abstract:

Ashe juniper is a problematic plant in central Texas due to rapid expansion causing the alteration of rangeland plant communities and hydrological landscapes. The thermochemical conversion of Ashe juniper into biochar with specific properties is a promising way for valorization of Ashe juniper. This study presents the effects of temperature (350, 450 and 520°C) and vacuum pressure (0.09, 0.7 and 3 kPa) on physical and chemical properties of biochar produced by vacuum pyrolysis from Ashe juniper. We found that temperature had a major effect on the degree of carbonization, surface functionality, as well as the O/C and H/C ratios regardless of vacuum pressure. The severity of vacuum pressure effects on product properties was found to depend on pyrolysis temperature. It was also confirmed that the vacuum pressure was a crucial factor for controlling the surface area and porosity at each constant pyrolysis temperature. This work also demonstrated that the biochar from Ashe juniper under vacuum pyrolysis showed improved aromaticity and surface area over the reported biochar derived from other raw biomass. Biochar produced at 450 and 520°C showed potential application as an adsorbent, a catalyst support and a carbon sequestration agent. Methylene blue (MB) adsorption test confirmed that biochar produced at 450°C and 0.7 kPa showed potential as an adsorbent.

495552 Renewable Transportation Biofuel Converted from Wet Biowaste Via Hydrothermal Liquefaction

Abstract:

This study demonstrates a proof-of-concept in the production of high quality renewable biofuel from wet biowaste via hydrothermal liquefaction (HTL). Distillation was employed to effectively separate the biocrude oil converted from swine manure (SW), food processing waste (FPW), and *Spirulina platensis* (SP) via HTL into different fractions. Distillation curves of different types of HTL biocrude oil were reported. Physicochemical characterizations, including viscosity, elemental test, chemical compositions, and acidity, were conducted on distillates separated from different feedstocks. SW-, FPW-, and SP-derived biocrude respectively contains 15 wt.% , 56 wt.%, and 15 wt.% distillates with heating values of 43-46 MJ/kg and alkanes with carbon numbers ranging from C8 to C18. Compared to the distillates from SW- and SP-derived biocrude oil, the distillates from FPW-derived biocrude demonstrate the closest energy content to petroleum diesel, though this type of distillates contained an excessively high acidity that need to be reduced from 35.3 mg KOH/g to â‰¤ 3 mg/g (the requirements suggested by the ASTM standard). An orthogonal array design of esterification experiment was performed to optimize the reaction temperature (50-70Â°C), reaction time (0.5h-6h), catalysts concentration (0.5 wt.%-2 wt.%), and the molar ratio of FPW-distillates to methanol (1:5-1:15), for achieving the lowest acidity. Compared to other available methods to upgrade HTL biocrude oil, the integrative upgrading approach proposed by this study (distillation plus esterification) demonstrates a competitive energy consumption ratio (0.03-0.06) to zeolite cracking (0.07), supercritical fluid (SCF) treatment (0.17), and hydrotreating (0.24). Moreover, the reaction severity (R_0) of the upgrading approach used in this study (with log R_0 of 5.9-9.5) is much lower than those of zeolite cracking (with log R_0 of 11.0), SCF treatment (with log R_0 of 10.6), and hydrotreating (with log R_0 of 11.3). Finally, the fuel specification analysis and engine test were conducted with the drop-in biodiesel, which was prepared with 10 vol.% (HTL10) and 20 vol.% (HTL20) upgraded distillates and 80-90 vol.% petroleum diesel. HTL10 and HTL20 exhibited a qualified Cetane number (>40 min), lubricity (<520 μm), and oxidation stability (>6 hr), as well as a comparable viscosity (0.2%-19% lower) and net heat of combustion (3%-4% lower) to those of petroleum diesel. Further, diesel engine tests demonstrated that HTL10 can lead to a superior power output (8% higher) and lower emissions of NO_x (3-7%), CO (1-44%), CO_2 (1-4%), and unburned hydrocarbons (10-21%). The present study showcases an energy-efficient and technically cohesive approach to produce renewable high-quality drop-in biofuels for demanding transport applications.

502978 Co-Gasification of Woody Biomass and Chicken Manure

Abstract:

In this work, the potential of gasification as a means of disposal of chicken manure, by converting this animal waste at hen layer farms into clean energy, was studied. Gasification experiments were conducted in a downdraft gasifier and the quality of producer gas generated was analysed. While gasification of pure wood waste resulted in producer gas with a lower heating value (LHV) of about 4.68 MJ/Nm^3, co-gasification of 30 wt% chicken manure and 70 wt% wood waste also generated producer gas of comparable quality, with LHV of 5.23 MJ/Nm^3. From this observation, chicken manure could be a suitable feedstock for co-gasification with woody biomass. In the second part of the study, the solid residue generated â€" biochar â€" was used to study its potential in water treatment, in particular the removal of emerging contaminant, as biochar is known for its superb adsorptive properties. Our results demonstrated that artificial sweeteners were successfully removed by the biochar. By studying the relation of pH, zeta potential, and adsorption efficiency, it was also predicted that the mechanism governing the adsorptive removal was likely electrostatic interaction and specific interaction. Last but not least, a cost-benefit analysis was conducted using a Monte Carlo simulation model to evaluate the economic feasibility of deploying a gasification system in a hen layer farm from business point of view.

Simulation Study of Single-channel Closed Cold Compression Molding for Straw Biomass using ANSYS Workbench

Xiaofei Xin[1,2], *Zhiwei Wang*[1,2,*], *Tingzhou Lei*[2,*], *Miao Yang*[1,2], *Xiaofeng He*[1,2], *Qian Guan*[1,2], *Gaofeng Chen*[1,2], *Tian Qi*[1,3], *Xueqin Li*[1,2], *Yifeng Wu*[1,2], *Yi Gao*[1,2]

(1Energy Research Institute Co. Ltd., Henan Academy of Sciences, Zhengzhou 450008, China

2Henan Key Lab. of Biomass Energy, Zhengzhou 450008, China

3Tianjin University, Tianjin 300072, China)

ABSTRACT: To simulate and analyze the creep process during the cold compression of agricultural and forestry waste and other biomass, we combined the biomass chemistry principles of elastic-plastic theory and contact analysis to establish a three-dimensional simulation model using SOLIDWORKS software. In addition, a single-channel closed compression process was simulated using ANSYS Workbench software. The displacement, equivalent stress, shear stress, X-direction strain, and elastic-plastic strain change laws of straw biomass during cold compression and the equivalent stress of the mold were determined. Furthermore, the creep law of the cold compression densification process was theoretically revealed. These results provide a basis for the formation mechanism and mold design scheme of the tested process.

KEYWORDS: ANSYS Workbench; biomass; cold compression; simulation

1. Introduction

China produces a large amount of agricultural and forest waste each year, which can be converted into about 500 million tons of standard coal[1]. Biomass energy is the only renewable energy that can be easily stored and transported. Given its uneven distribution and relatively low energy density, many physical conversion technologies, such as biomass densification molding, have been adopted domestically and internationally[2]. Molding technology transforms loose biomass into solid compression-molded fuel, which can improve the density of the raw material up to 1 t/m^3. Molding can improve the combustion characteristics and can be used as a pretreatment for biomass gas fuels, liquid fuels, and other industrial products[3]. It is an important technique for the comprehensive realization of large-scale biomass resource use. Cold compression molding has the advantages of high adaptability to raw materials and water contents of materials, and low power consumption, and has been widely promoted in the industry[4].

A number of studies have examined the cold compression molding process. Dong et al. [5] and Wang et al. [6] determined the change laws of the load during the material densification process using sawdust as the research material and a finite element simulation of the extrusion process, which was performed using ANSYS software. Combined with information on the typical structure of the biological properties, Sun et al. [7] simulated and analyzed the relationship between pressure deformation and pressure compression density during the extrusion stage. Based on a theoretical study of nonlinear problems, Hu and colleagues [8-10] studied the influence of the structural parameters of different molds on the compression process. Sun et al. [11] determined the cohesion and internal friction angle of straw powder using a triaxial compression test and analyzed the compression process of straw powder with thermal coupling. The ANSYS Workbench platform has been developed rapidly in recent years, and features a user-friendly interface and good three-dimensional (3D) effects. In this study, the biomass cold compression molding process was simulated using ANSYS Workbench based on the theories of elastic-plastic mechanics and contact mechanics. This analysis of the displacement, stress, strain, shear stress, and elastic-plastic strain laws of biological material under compression and the stress of the mold provide a basis for the formation mechanism and mold design scheme of the tested process.

2. Constitutive Relationship of the Biomass Elasto-plastic Model

2.1 Yield Criterion

The biomass compression process is divided into the loose stage, transition stage, compression stage, and push stage. A simple elastic model cannot satisfy the simulation demand of analyzing the complexity of the compression process, and the plastic constitutive equation (stress-strain relationship) of the material is nonlinear. The critical value of the material entering the plastic state is the yield limit under uniaxial tension and compression[3]. The stress-strain relationship of elastic phase materials obeys the generalized Hooke's law $\sigma_{i,j} \sim \varepsilon_{i,j}$,

$$\varepsilon_x = \frac{1}{E}\left[\sigma_x - \upsilon\left(\sigma_y + \sigma_z\right)\right], \gamma_{yz} = \frac{\tau_{yz}}{G}$$
$$\varepsilon_y = \frac{1}{E}\left[\sigma_y - \upsilon\left(\sigma_z + \sigma_x\right)\right], \gamma_{zx} = \frac{\tau_{zx}}{G} \quad (1)$$
$$\varepsilon_z = \frac{1}{E}\left[\sigma_z - \upsilon\left(\sigma_x + \sigma_y\right)\right], \gamma_{xy} = \frac{\tau_{xy}}{G}$$

The yield condition formula can be written as follows:

$$F(\sigma_{i,j}) = 0 \quad (2)$$

Equation (2) represents the yield surface around the origin. When $F(\sigma_{i,j}) < 0$, the stress point should be within the surface and the material is in the elastic state. When the material has satisfied $F(\sigma_{i,j}) = 0$, it begins to yield and enters the plastic state. Assuming that the material is initially isotropic, the yield condition is independent of the coordinate axis and can be written as the principal stress function or stress invariant function:

$$F(\sigma_1, \sigma_2, \sigma_3) = 0 \text{ or } f\left(J_1, J_2, J_3\right) = 0 \quad (3)$$

When the Mises yield condition considers the influence of the intermediate principal stress, when Equation (4) is satisfied, the material will enter the yield state, and C is constant in the formula.

$$J_2' = C \quad (4)$$

2.2 Hardening Criterion

Because the isotropic model is considered to be similar to the expansion of the loading surface within a space, the material is considered to maintain its isotropic properties after entering the plastic deformation state only under the conditions of minimal deformation and an unchanged stress ratio. Because of the internal structure of plant cells, biomass material is predominantly anisotropic. In this study, straw biomass was considered to follow the kinematic hardening rule, which is applied to large plastic deformation where the stress changes repeatedly. Under the uniaxial stress state, the Bauschinger effect decreases the yield stress in the opposite direction, and the total elastic range remains constant. Assuming that the size of the yield surface remains constant and only moves in the yield direction, the initial material will no longer remain isotropic after yielding. At the inner center of the yield surface, the coordinates are zero during the initial yield, and the loading surface is expressed as follows:

$$f\left(\sigma_{ij}, H_a\right) = f\left(\sigma_{ij} - \alpha_{ij}\right) = f^*\left(\sigma_{ij} - \alpha_{ij}\right) - k = 0 \tag{5}$$

In the formula, $f^*\left(\sigma_{ij}\right) - k = 0$ is the initial yield surface and moves with α_{ij}, which is the moving tensor, and depends on the amount of plastic deformation. When $\alpha_{ij} = c\varepsilon_{ij}^p$, for linear hardening, C is considered as a constant.

3. ANSYS Workbench Finite Element Analysis

Figure 1 presents a flow chart of the basic analysis of the biomass compression molding simulation process.

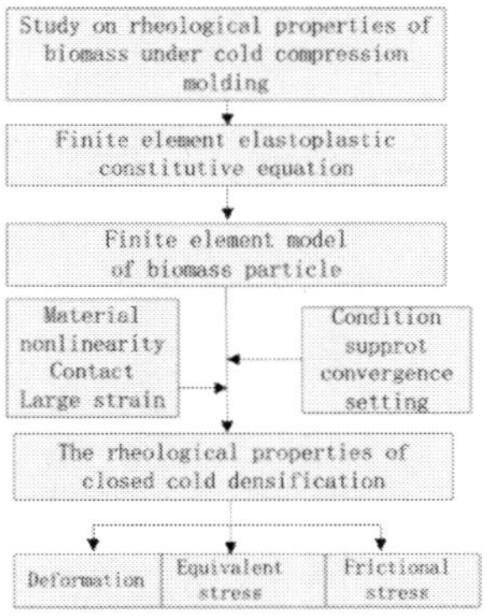

Fig. 1 Analysis flow chart of biomass compression

3.1 Self-Made Mold Design

A 3D simulation model was established using SOLIDWORKS software, where the cone angle of the mold was 60°. The biomass materials were compressed from the upper entrance of the single channel and compressed in the biomass formation chamber. The 3D model was directly imported into ANSYS Workbench and checked for data loss.

3.2 Meshing and Material Settings of the 3D Model

Since the mold was cylindrical, one-quarter of the mold was used as a simplified analysis model. The model was divided into three parts, the mold, biomass, and pressure bar, which was constructed of structural steel, so that the small section could be intercepted. Considering the nonlinearity of the material, the contact nonlinear problem, and the large strain problem of the compression process, the high-order SOLID186 element type and the structure form of meshing method were adopted. The meshing is shown in Figure 2; the model was divided into 1,478 units and 8,235 nodes.

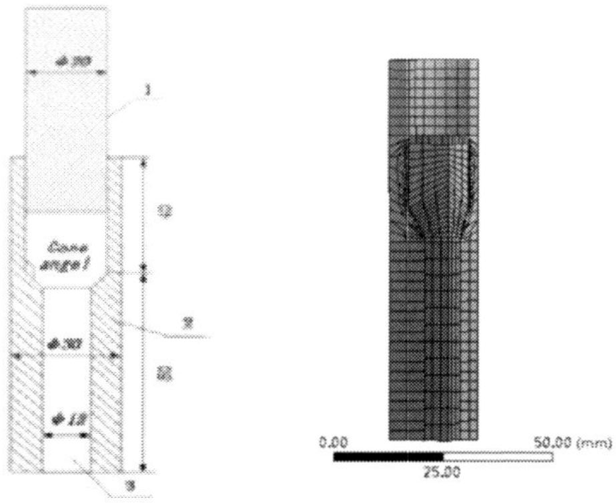

Fig. 2 Meshing of the 3D Models

The biomass, mold, and pressure bar parameters were each set in Engineering Data, and the material data were set as listed in Table 1.

Table 1 Setting parameters of the materials

Material type	Parameter	Setting
Straw/ Biomass	Density (ρ /kg • m^{-3})	330
	Poisson's ratio (υ)	0.35
	Elasticity modulus (E/GPa)	160
	Yield strength (MPa)	90
Stainless steel/Mold	Density (ρ /kg • m^{-3})	7750
	Poisson's ratio (υ)	0.28
	Elasticity modulus (E/GPa)	210
Structural steel/	Density (ρ /kg • m^{-3})	7850
	Poisson's ratio (υ)	0.3

Pressure bar	Elasticity modulus (E/GPa)	200

3.3 Contact Settings

During the cold compression densification process, a highly nonlinear problem existed at the point of contact between the bar and biomass as well as the biomass and mold. SOLIDWORKS formed the contact pairs automatically. The contact setting parameters were defined in the Connections section. Considering that the gap of the elastic modulus between the three components was large, the entire biomass body was defined as the contact body, and surface contact was defined as rigid-soft contact. The contact types were defined as Bonded and Frictional, the friction coefficient was defined as 0.2, and the behavior was defined as Asymmetric. The augmented Lagrange formulation, which is applicable to large strain problems and nonlinear problems, was used and the normal stiffness factor was defined as 0.1[12].

3.4 Constraints and Loads Settings and Solve Settings

The bottoms of the mold and biomass were provided with Fixed Support, and the dividing surfaces were provided with Displacement in the corresponding direction set to zero. Pressure bar displacement was applied to restrict the Displacement. Considering the complexity of the compression process, the full Newton-Raphson option, line search in the nonlinear option, and large deflection option were applied to promote the convergence of the simulation results[13].

4 Results and Analysis

4.1 Stress and Strain Distribution of Biomass

Figure 3 shows a flow diagram of the four stages in the biomass compression phase. As can be seen in Figure 3, as compression displacement increased gradually, the biomass compression state gradually moved downward, and the compressive deformation of the material maintained a degree of hierarchy from top to bottom. The initial stage was uniform and the flow state changed after entering the variable cross-section. The flow state of the middle layer of biomass was markedly faster than that of the outer layer, and the compression path was funnel-shaped and intensified inward. Because of the pressure, relatively large shear stress and friction force formed at the cross-section, and the material of the biomass was compacted under the various forces, resulting in a degree of non-uniformity of the density.

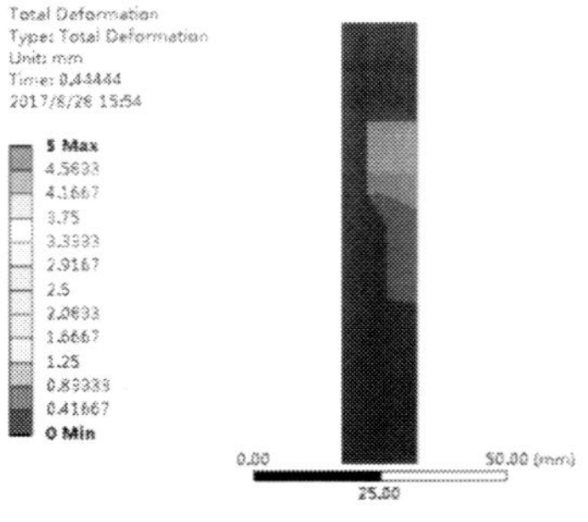

a. t = 0.45 s

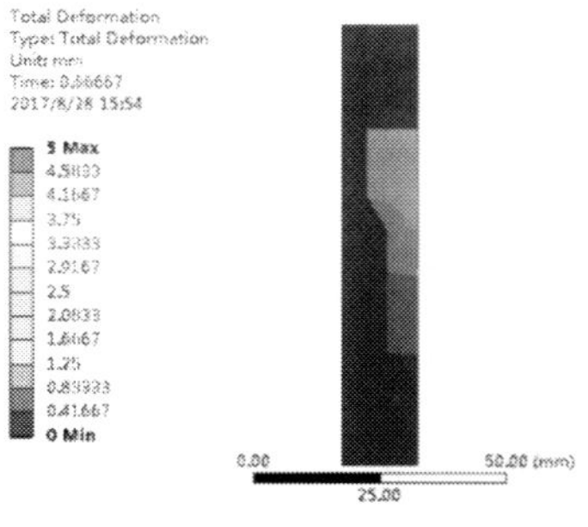

b. t = 0.67 s

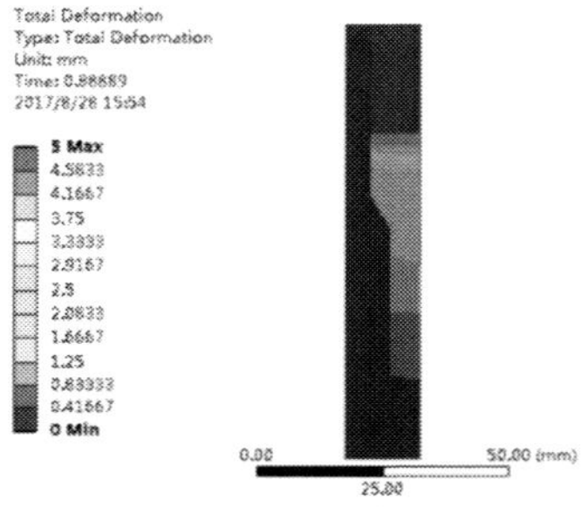

c. t = 0.89 s

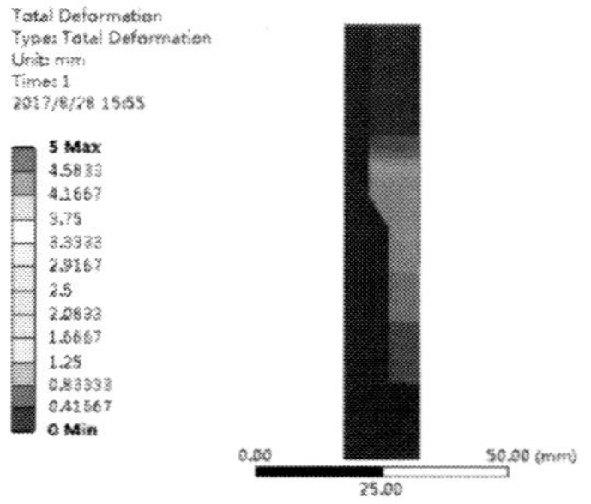

d. t = 1 s

Fig. 3 Flow diagram of biomass at different stages of compression

The path of direction 1-2 on the variable cross-section was selected for further analysis (Fig. 4). As the biomass passed through the variable cross-section, the displacement distribution curve changed markedly [Fig. 4(b)]. As the variable cross-section moved downward, the biomass deformation increased rapidly. Deformation remained roughly stable when the length reached 3 mm but declined around 5 mm. The variation of the equivalent stress curve was relatively stable, and the range of stress was 75–90 MPa.

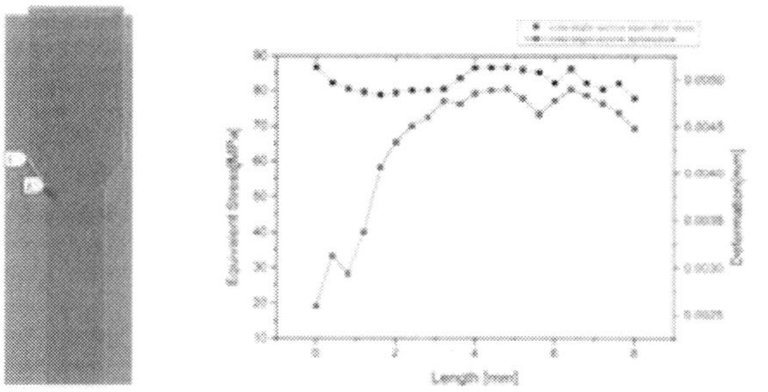

a. Path of the cross-section (1-2) b. Displacement and stress distribution

Fig. 4 Variations in the cross-section

To further assess the stress and strain of the biomass, three paths were selected as biomass cross-sections, inlet section 1, middle section 2, and outlet section 3. The deformation and equivalent stress distribution curves are shown in Figures 5 and. 6.

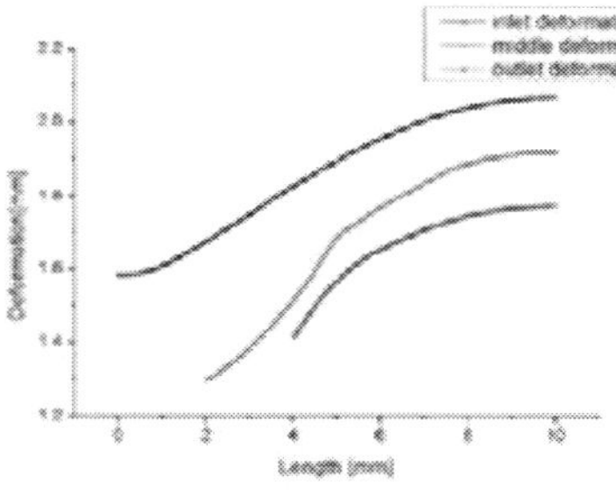

Fig. 5 Path deformation distribution

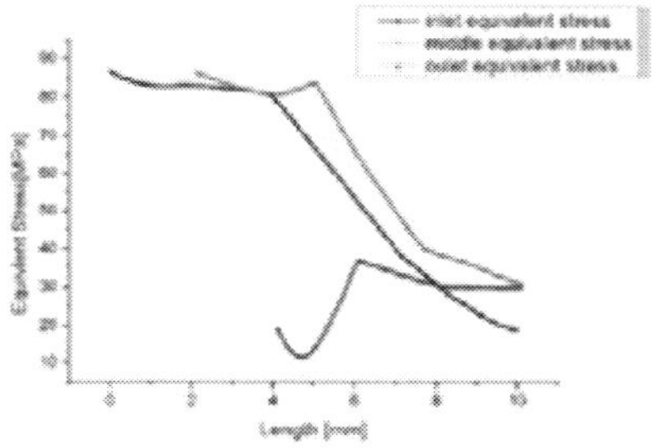

Fig. 6 Path equivalent stress distribution

The biomass was compressed into the variable cross-section, where the middle area had the largest displacement and the complete section flow state was parabolic (Fig. 5). At middle section 2, raw material near the inner wall of the mold was in a state of stagnation under bar pressure, frictional force, and cone force, and elastic deformation occurred. The minimum equivalent stress area was in the middle of the biomass material, which gradually increased toward the mold wall (Fig. 6). The equivalent stress reached a maximum force of 80–90 MPa near the wall under the action of bar pressure and friction, which was consistent with the minimal displacement. Plastic deformation occurred at the near inner wall of outlet section 3, resulting in reverse shear force that led to a marked decrease in equivalent stress. With the downward compression of the material, the displacement gap between the intermediate and both sides gradually narrowed, and the flow state tended to stabilize. In addition, the equivalent stress distribution became more stable.

4.2 XY Plane Shear Stress and X-direction Strain of Biomass

Due to the pressure, the shear stress at the variable cross-section was relatively large, and the XY plane shear stress of the biomass changed over time (Fig. 7). Shear stress first occurred at the inflection points of the entrance and exit at the variable cross-section. As the compression process progressed, the shear stress gradually increased to 49 MPa, the two regions gradually expanded, and a layered force field formed. Near outlet section 3, a reverse shear stress of 20 MPa appeared in the middle of the material by the mold wall. The raw biomass material was further broken down under the joint action of shear stress and compressive stress, and the section of the material near the inner area of the mold was gradually broken down and gathered toward the center for bonding.

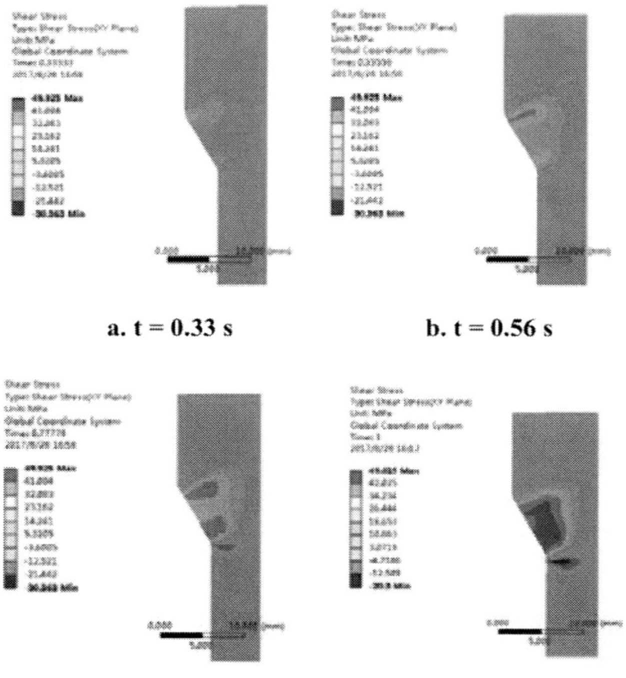

a. t = 0.33 s **b. t = 0.56 s**

c. t = 0.78 s **d. t = 1 s**

Fig. 7 XY plane shear stress distribution

The X-direction strain began at the inflection point of the cross-section entrance, and gradually expanded downward and toward the middle region (Fig. 8). The X-direction strain was closely related to the shear stress of the biomass. As shear stress gradually increased, the material near the inner wall of the mold was constantly squeezed toward the middle region.

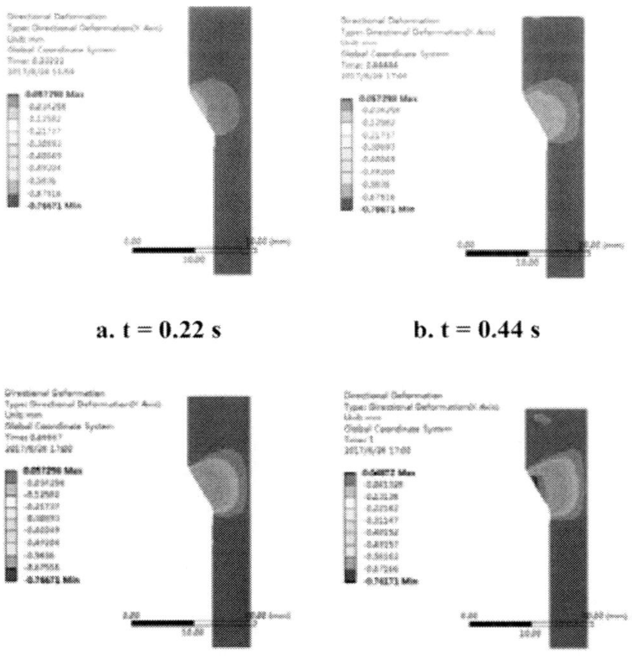

a. t = 0.22 s **b. t = 0.44 s**

c. t = 0.67 s **d. t = 1 s**

Fig. 8 X-direction strain distribution

4.3 Elastic Strain and Plastic Strain

Figure 9 shows the equivalent elastic strain and equivalent plastic strain of biomass. Compared with the above analysis, the biomass inlet material was mainly in the elastic stage, and elastic changes mainly occurred in the middle area of the material. Plastic strain mainly occurred at the contact point of the inlet and outlet corner of the variable cross-section, as well as in other contact areas.

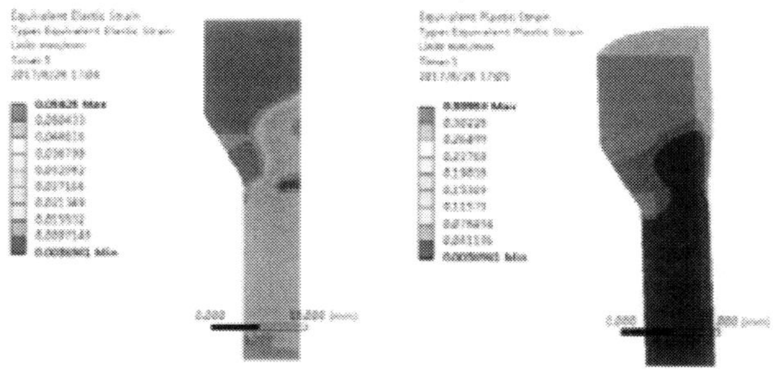

a. Equivalent elastic strain **b. Equivalent plastic strain**

Fig. 9 Elastic and plastic strain distribution

4.4 Equivalent Stress of the Mold

Figure 10 presents the equivalent stress distribution of the 60° mold. The equivalent stress was mainly concentrated on the inside of the variable cross-section, where the stress concentration was much larger than that of the other parts of the mold. If the angle of the cone changed, the equivalent stress of the mold would change substantially.

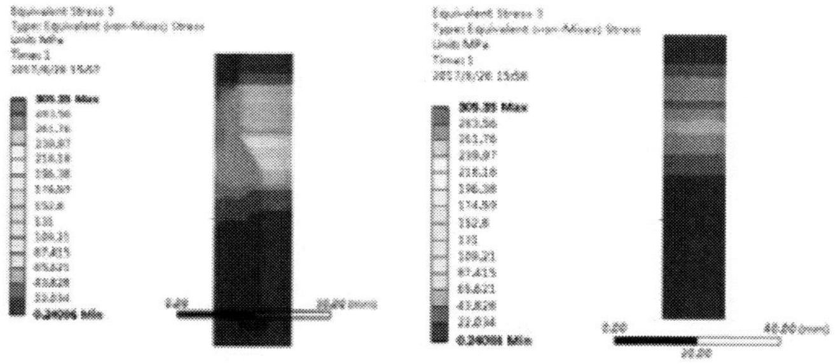

a. Rear view of the mold **b. Front view of the mold**

Fig. 10 Equivalent stress distribution of the mold

4.5 Contact Analysis

In the initial stage of compression, the frictional stress between the biomass material and the mold was in an elastic relationship, indicating that the compressive force of the bar to the biomass obeyed Hooke's law over time, mainly in the elastic stage. As the pressure increased and the compression process progressed into the compression stage, the frictional stress increased accordingly. As a result of elastic and plastic deformation, the friction appeared to fluctuate slightly, but still exhibited an overall increasing trend.

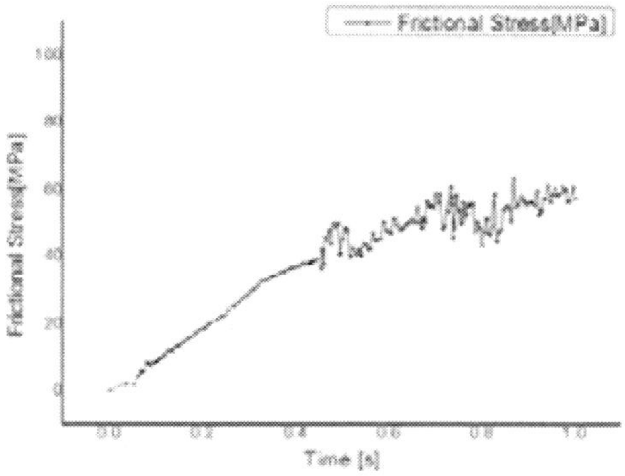

Fig. 11 Frictional stress distribution between the biomass and the mold

5 Conclusions

We established an accurate 3D model of the characteristics of biomass undergoing compression and the compression process. The closed cold compression process was simulated with ANSYS Workbench, and the flow characteristics of the creep process were analyzed. The major results are summarized as follows.

(1) In the single-channel closed cold compression molding process of biomass, the pressure was transferred from top to bottom, and the flow of material showed a degree of change. Under the joint action of various forces, the flow in the middle region was faster than that in the inner wall of the mold, and assumed a funnel-shaped flow. The displacement and equivalent stress distribution curves of different sections of the mold were compared, and the displacement deviation of the middle region and the area on both sides gradually decreased and stabilized, and the equivalent stress distribution tended to become uniform.

(2) Shear stress first occurred at the inlet and outlet inflection points of the variable cross-section, which gradually expanded from the center of the two regions. Meanwhile, the center material exhibited reverse shear stress near outlet section 3. The strain in the X-direction was closely related to the shear stress of the biomass. As the shear stress gradually increased, the material near the inner wall of the mold was squeezed toward the middle region. Elastic strain mainly occurred at the entrance of the raw material and in the middle area, and plastic strain mainly occurred in the contact region of the inlet and outlet inflection point.

(3) The stress concentration at the inner cross-section of the mold was much larger than that of other areas, and the lateral area experienced transition stress. The frictional stress between the mold

and the biomass predominantly obeyed Hooke's law, and entered the plastic state and fluctuated under the action of pressure force and variable section force.

Acknowledgments: This study was supported by the National Natural Science Foundation of China (grant no. 51476050) and the Henan Science and Technology Tackling Project (grant nos. 172102410030, 162102210390, and 152102310392).

References:

1. Zh. Yuan, Ch. Wu, and L. Ma, Principles and Technologies of Bioenergy Utilization (Chemical Industry Press, 2016).

2. S.J. Liu, L.P. Abrahamson, and G.M. Scott, Biomass Bioenergy 39, 1 (2012).

3. Z. Wang, T. Lei, M. Yang, Z. Li, T. Qi, X. Xin, X. He, A. Ajayebi, and X. Yan, Appl. Energy 192, 83 (2017).

4. R. Wang, Plastic Mechanics Foundation (Peking University Press, 1982).

5. Y.P. Dong, M.W. Gao, and Q.X. Sun, J. Shandong Univ. Sci. 35, 9 (2005).

6. W. Wang, Y. Dong, S. Zhou, H. Sun, and W. Ding, J. Agric. Mech. Res. (2011).

7. Q.X. Sun, R.J. Zhang, and Y.P. Dong, Trans. Chin. Soc. Agric. Mach. 130 (2009).

8. J.J. Hu, G.Y. Xu, J.W. Liu, T.Z. Lei, and S.Q. Shen, Bioresources 5, 2447 (2010).

9. J.J. Hu, Straw Pellet Fuel Cold Molding by Compression: Experimental Study and Numerical Simulation, Dalian University of Technology, 2008.

10. J. Hu, T. Lei, X. He, Z. Li, J. Liu, Z. Wang, and J. Zhu, Acta Energiae Solaris Sin. 29, 241 (2008).

11 Y. Sun, X. Bai, and Y. Li, J. Agric. Mech. Res. 237 (2016).

12 Y. Dong and J. Gao, Acta Energiae Solaris Sin. 31, 661 (2010).

13 R.X. Zhao, Study on Optimization of Biomass Fuel Pressure Formation Process Parameter, North China University of Science and Technology, 2016.

The English in this document has been checked by at least two professional editors, both native speakers of English. For a certificate, please see:

http://www.textcheck.com/certificate/M7L8iq

489620 High-Performance Magnetic Activated Carbon from Solid Waste from Lignin Conversion Processes. 1. Their Use As Adsorbents for CO_2

Abstract:

Wood is a major source of raw materials for renewable fuels, chemicals and materials. Depending on its type, lignin can account for roughly one-third of the mass. It is considered a waste product in most of the processes that convert wood into valuable products. In the paper and pulp industry, where most of the lignin is available, 98 % is combusted to supply the processes with energy and low-grade heat[1]. This combustion is wasting potential value of lignin, both in an economical aspect and in the potential replacement of fuels, materials and chemicals based on fossil resources. Lignin can be used to produce value-added and renewable products in lignin-to-liquid processes. In these processes, lignin is depolymerized to form bio-oil containing various useful organic molecules for use in chemicals, fuels and materials[2]. Integrating such processes can be important in future biorefineries, where renewable resources are converted in to useful products. However, these processes produce large amounts solid by-products which, so far, has not found any constructive use. In the biorefinery concept, it is imperative to maximize the use of the renewable raw materials, both for the sustainability of the process and to generate enough added values to make the sustainable processes economically viable. In this work, we used the solid by-products of lignin-to-liquid processes to produce highly porous and magnetic activated carbons for use in carbon dioxide separation processes. Activated carbons have many commercial applications within separation, deodorization, purification and catalysis[3]. However, the majority of activated carbons are produced from fossil resources[4] and there is a need to replace these fossil-based materials with renewable alternatives. In carbon dioxide separation processes, activated carbons can potentially lower the carbon dioxide capture cost, which is a problem for CCS (carbon capture and storage) systems[5]. The general advantages of activated carbons compared to other solid sorbents are high adsorption capacities, stable adsorption/desorption cycling and low production costs. To produce the magnetic activated carbons in this work, Spruce and Eucalyptus lignin was heated at 380 °C with water, formic acid and a catalyst, the solid by-product of this bio-oil producing process was mixed with an aqueous solution of potassium hydroxide, heated at 200 °C for 5 hours and activated under nitrogen gas flow for 4 hours at 700-800 °C. The carbons activated at 700 °C exhibited carbon dioxide adsorption capacity of up to 6.0 mmol/g at 273K and 1 atm and very high ultramicropore volumes. The carbons activated at 800 °C exhibited BET specific surface areas, measured by nitrogen adsorption at 77K, of up to 2875 m^2/g. The activated carbon based on Eucalyptus lignin and activated at 700°C showed an isosteric heat of CO_2 adsorption was 27-30 kJ/mol, a simplified description indicated a CO_2-over-N_2selectivity of 14, while a full IAST prediction indicated a selectivity of between 3 and 4, and very good cyclic performance. All the activated carbons exhibited soft magnetic behaviour due to embedded nanoparticles of iron and magnetite. We show that the activation temperature does not only alter the porosity of the carbons but also the textual properties. The combination of the highly porous nature, the magnetic properties and the option to optimize the porosity through the

activation temperature makes these materials highly interesting for many other potential applications outside CCS as well. This work has been published elsewhere[6].

[1] R.J.A. Gosselink, E. De Jong, B. Guran, A. Abächerli, Co-ordination network for lignin - Standardisation, production and applications adapted to market requirements (EUROLIGNIN), Ind. Crops Prod. 20 (2004) 121–129. doi:10.1016/j.indcrop.2004.04.015.

[2] M. Kleinert, T. Barth, Towards a lignincellulosic biorefinery: Direct one-step conversion of lignin to hydrogen-enriched biofuel, Energy and Fuels. 22 (2008) 1371–1379. doi:10.1021/ef700631w.

[3] R.C. Bansal, M. Goyal, Activated carbon adsorption, CRC Press, 2005.

[4] A. Ahmadpour, D.D. Do, The preparation of active carbons from coal by chemical and physical activation, Carbon N. Y. 34 (1996) 471–479. doi:10.1016/0008-6223(95)00204-9.

[5] V. Presser, J. McDonough, S.-H. Yeon, Y. Gogotsi, Effect of pore size on carbon dioxide sorption by carbide derived carbon, Energy Environ. Sci. 4 (2011) 3059. doi:10.1039/c1ee01176f.

[6] W. Hao, F. Björnerbäck, Y. Trushkina, Mikel O. Bengoechea, G. Salazar-Alvarez, T. Tarth, N. Hedin, High-Performance Magnetic Activated Carbon from Solid Waste from Lignin Conversion processes. 1. Their Use As Adsorbents for CO2, ACS Sustainable Chem. Eng., 2017, 5 (4), pp 3087–3095, doi: 10.1021/acssuschemeng.6b02795

501752 Aromatic Ring Opening of Lignin for Fuel and Chemical Production

Abstract:

Lignin consists approxiamately one third amount of carbon in lignocellulosic biomass. Emerging biorefinery will generate enormous amount of lignin. The high C/H ratio made lignin a promising feedstock for fuel and value-added chemical production. Unfortunately, existing lignin to liquid fuel conversion technologies rely on hydrogenation and hydrodeoxygenation, processes with higher energy requirements and excessive usage of external hydrogen that make these technologies uncompetitive with petroleum-based aviation fuel production. In addition, these processes yield aromatic or cyclic hydrocarbons as the main products; the difficulty in manipulating the resulting product chain length significantly limits their application. Selectively converting lignin to open chain hydrocarbon fuel is apparently an appealing idea needs input of efforts.

In this talk, we will present a new chemistry pathway allows production of open-chain hydrocarbon from lignin. The chemistry will generate precursors for fuel and high value chemicals simultaneously. This work also aiming provide new perspectives towards the understanding and design of lignin sequential depolymerization and aromatic ring cleavage.

495705 Aqueous Lignin Purification with Hot Acids: Cleaning, Fractionating, and Solvating Lignin for Materials Applications

Abstract:

Thies and co-workers [1,2] have discovered that when raw lignins are combined with hot acetic acidâ€“water (AA/H2O) mixtures, two liquid phases are formed: a solvent phase into which are extracted the metal salts and other impurities and an ultrapure, solvated lignin phase that is the desired product. Purities of less than 20 ppm Na and <100 ppm total metals can be achieved in this liquefied lignin phase, as the metal salts readily diffuse out of the solvated lignin and into the solvent. Furthermore, the AA/H2O solvent composition can be â€œtunedâ€ to partition the raw lignin feed between the solvent and lignin phases according to molecular weight (MW). Finally, because the ultrapure lignin phase is solvated, it has the potential to be directly processed into useful products. No other lignin-treatment method reported in the literature exhibit the ability to simultaneously clean, fractionate, and solvate lignins without washing. We call the process ALPHA, which stands for Aqueous Lignin Purification using Hot Acids. A low-ash, raw Kraft lignin has been purified via ALPHA, dry-spun from the solvated form into lignin fibers, and then processed into the desired final carbon fibers. As a reference and first step, the raw lignin was solvated by ALPHA but was not fractionated or cleaned before conversion into carbon fibers. In succeeding experiments, 50/50, 65/35, and 67/33 AA/H2O solvent mixtures were used to clean, solvate, and fractionate the raw lignin so that a highest 50% MW cut, a highest 25% MW cut, and a highest 10% MW cut were isolated. Conversion of these cuts into carbon fibers showed consistent improvements in fiber properties with increasing number average molecular weight of the lignin.

1. Klett, A. S.; Payne, A. M.; Thies, M. C. Continuous-Flow Process for the Purification and Fractionation of Alkali and Organosolv Lignins. *ACS Sustainable Chem. Eng.*, **2016**, *4*, 6689-6694.
2. Klett, A. S.; Chappell, P. V.; Thies, M. C. Recovering ultraclean lignins of controlled molecular weight from Kraft black-liquor lignins. *Chem. Comm.* **2015**, *51*, 12855-12858.

497066 Lignin Value Prior to Pulping â€“ Analyzing Feasibility

Abstract:

Organic solvents can facilitate the production of chemicals, fuels, nanocellulose, and pulp from lignocellulosic biomass. For these applications, the majority of research focuses on how specific organic solvents isolate and alter native cellulose. Specifically, the largest body of research focuses on the production of cellulosic ethanol. The present work seeks to leverage an understanding of the pulp and paper industry, the biggest consumer of cellulose and lignin. Over a billion tons of wood are processed annually, and marginal improvements can have large economic impact. In this work, we describe a process that employs an organic solvent to remove lignin from biomass while maintaining cellulosic integrity for pulp production. This next-generation pulping system, called lignin value prior to pulping (LVPP), provides an un-sulfonated lignin stream and improves the operating economics of a pulp mill. Through a technoeconomic analysis, we describe the LVPP implementation of organic solvents like alkylene carbonates and alcohols. This analysis provides a minimum lignin-selling price (MLSP), and highlights the feasibility of an LVPP process.

500655 Surface Structure Patterning for Fabricating Non-Fluorinated Super-Hydrophobic Cellulosic Membrane

Abstract:

Superhydrophobic surface have attracted great interesting in scientific and industry. Surface pattern-featured nano-cellulose membranes were fabricated by a simple pressing process. After chemical modified by non-fluorine chemicals, such as AKD, wax and isocyanates, the nano-cellulose membrane shows water contact angles greater than 150°. The micro-sized pattern and nano-sized cellulose fibers create a hierarchical structure that improves the superhydrophobic properties. SEM, optical microscopy and profile simulation were used to evaluate characters in the fiber network and hierarchical scale structures. Wetting testing experiments shows water droplets can roll off from the cellulose membrane, which may be used as liquid packaging materials. The process used in this superhydrophobic cellulose membrane fabrication is simple, cost effective and amenable to scale-up.

501268 Photonic Cellulose Nanocrystal (CNC) Coatings

Abstract:

Retention of cellulose nanocrystal (CNC) dispersionsâ€™ cholesteric microstructure in dried films can enable photonic materials that exhibit iridescence and selective reflection of circularly polarized light. Such materials would require uniform planar anchoring of the cholesteric helix; they could be used for optical filters, mirrorless lasers, and optical sensors. However, nonuniformities in the initial dispersion and microstructural changes during film casting and drying have made it challenging to obtain uniform optical reflection over more than a few microns. In this research, investigation of CNC drying kinetics and surface anchoring has enabled uniform planar orientation of the CNC helix in coatings on a lateral scale of few hundred micrometres. Uniform planar orientation and the associated selective reflection, was confirmed by microspectrophotometry, polarized light microscopy, and scanning electron microscopy. In addition to providing a foundation for the development of CNC based optical materials, this study also provides new scientific insights into the helical defects reported in natural chiral nematic photonic structures such as those of *lomaptera* beetles and *pollia* fruits.

501694 Study Pyrolysis Products of Live and Dead Shrub Fuels from the Forest in the Southeastern Unites States

Abstract:

Prescribed fire, also referred to as controlled burning, is an effective way to reduce accumulation of fuels in forests. Prescribed fire contributes in hazardous fuel reduction, pest insects and disease minimization, habitat improvement for threatened and endangered species, and site preparation for trees and plants growth. During prescribed burning, solid fuels undergo irreversible thermal degradation when exposed to high temperature. The endothermic thermal degradation process consists of two sequential steps: pyrolysis and combustion. Pyrolysis is the first chemical reaction which occurs in the burning of some fuels like biomass, plants, wood, paper, and cloth. In this research, a flat-flame burner (FFB) apparatus was used to study pyrolysis of live and dead shrub fuels from the forests in the Southeastern United States. The major products of pyrolysis then analyzed using gas chromatography mass spectrometry technique, and the products classified as non-condensable gases (e.g., CO, CO_2, H_2O, and H_2), condensable gases (tars), solid residue (char), and mineral ash.

504637 The Study of the Feasibility of Joint Production of Jet Fuel and Bioethanol from Biomass

Abstract:

The cost of pretreatment and the inhibitors from the lignin and polysaccharide decomposed hinder the process of bio-ethanol commercialization. Develop a one-pot process which refinery the lignin first to jet fuel using thermochemistry ways, and then use the residual (contain cellulose main portion) to converse bio-ethanol was benefits a lot. The enzyme hydrolysis and fermentation efficency of the left substrate were analyzed. After apple wood catalyzed by Ni and Ru/C for jet-fuel production the surface area and accessibility of cellulose increased and the Cr. I of cellulose I was improved, which led to the improvement of the efficency of enzyme hydrolysis. After catalyzed by Ni, the substrate enzymatic hydrolysis increase from 5.8% (the raw apple wood) to 88%. Low Ni concentration (0.4g/l) was not inhibit the enzyme hydrolysis of the residual, but not the same effect of Ru/C. Ni was obvious inhibitor for yeast fermentation. Ru/C was not toxic to yeast fermentation in YPD. When carried on Q-SSCombF, the substrate (including catalyst) inhibited a lot of yeast fermentation even catalyzed by Ru/C. When separate hydrolysis and fermentation was carried, the ethanol conversion of the fermentation of the substrate catalytic by Ni was improved from 33% to 75% and the result of Ru/C catalytic substrate was increased from 25% to 73%. Carry on jet-fuel and bio ethanol co-production from biomass was feasible, the catalyst in substrates was the main point of the process.

504060 Continuous Distillation of Fast Pyrolysis Bio-Oils

Abstract:

Integration of pyrolysis-based biorefineries into petroleum refinery infrastructure will require novel strategies for bio-oil separations, including but not limited to distillation. Traditionally, distillation of fast pyrolysis bio-oils has been regarded as futile due to the thermal instability of the bio-oil. However, with developments in advanced pyrolysis techniques such as catalytic fast pyrolysis (CFP) and tail-gas reactive pyrolysis (TGRP) that reduce oxygen content and increase the thermal stability of bio-oils, distillation can proceed with significant yields that concentrate valuable chemicals into usable cuts. Still, little data on continuous separations has been offered. This talk will focus on preliminary data obtained from a continuous flash distillation system for pyrolysis bio-oil separations. The distillation system can be operated as either a one-stage or two-stage flash distillation. Based on our previous distillation studies, the first stage was operated at 120 °C, while the second stage can be operated between 250 and 300 °C. For one-stage distillation, the higher temperature set point was used. Yields and compositions of the fractions are analyzed and compared with results from batch experiments.

495467 Lignin Filers in Polylactic Acid Composite: The Effects of Impurities in Lignin and Lignin Type

Abstract:

Lignin filers in polylactic acid composite: The effects of impurities in lignin and lignin type

Yiwei Gao, Wangda Qu, Xianglan Bai

Lignin is second largest biopolymer on Earth. Lignin byproduct is abundantly available from paper and pulping industries, as well as emerging biorefineries. Due to its low cost, aromatic polymer structure and carbon rich nature, utilization of lignin for value-added products has been of great interest. For example, lignin filer is blended with other polymers to improve the properties of the composite materials, lower the product cost and enhance biodegradability. It is known that isolated lignin often contains considerable amount of inorganic impurities, and the impurities could affect the properties of lignin. Although the impurities are removable, purifying lignin can be a costly process. In the present study, the effect of impurities on the performance of lignin as the filer was studied by blending polylactic acid (PLA) with corn stover organosolv lignins with different amounts of impurities. The impurity content in the lignin was controlled by acid washing raw lignin with different numbers of washing and rinsing cycle. The PLA-lignin composites were characterized for their thermal, chemical and mechanical properties. Overall, blending corn stover lignin with PLA reduced the tensile strength of the composite but increased its modulus compared to pure PLA. When the ratio of lignin to PLA was fixed, using higher ash containing lignin resulted in the composite with better tensile strength. However, the composite with highest modulus was obtained with the lignin filer containing an intermediate amount of ash. For the lignin filer with fixed ash content, the composite containing 10% lignin filer had both highest tensile strength and modulus. It was also found that using raw lignin containing 7% of ash reduced thermal stability of the composite due to catalytic effect of the impurities for thermal decomposition. However, the lignins with lower ash contents did not significantly affect the thermal stability of the composites. A biorefinery produced hardwood lignin was also used as the filer in PLA to investigate the effect of lignin structure on the performance of lignin filer.

495477 Effect of Polyethylene Terephthalate in Producing Carbon Fiber from Pyrolytic Lignin

Abstract:

Pyrolytic lignin is a mixture of lignin-derived phenolic monomers and oligomers. Our previous study showed that repolymerized pyrolytic lignin could be used as a precursor to produce carbon fiber with enhanced mechanical properties. In the present work, pyrolytic lignin was mixed with different ratios of polyethylene terephthalate (PET) to extend the chain length of the precursor and control the molecular orientation. The FTIR spectra of the pyrolytic lignin and PET treated together at 200 °C were significantly different from the spectra of the physical mixture of individually treated pyrolytic lignin and PET, suggesting pyrolytic lignin and PET chemically reacted to form a new polymer structure. It is likely that pyrolytic lignin first caused chain scission of PET at the low temperature and then attached to the PET backbone through reacting with the carboxylic acid and vinyl groups on the PET fragments. The new precursors were further characterized using GPC, DSC, TGA, CHN and rheology analysis. The TGA analysis has shown that mixing pyrolytic lignin with 5% of PET increases carbon residue upon heating, thus potentially improving carbon fiber yield. The precursors were subjected to melt-spun at 200 ~ 210 °C and the spun fibers were oxidatively stabilized from room temperature to 250 °C using a heating rate of 0.3 °C/min. Subsequently, the stabilized fibers were carbonized at 1000 °C under inert environment. The produced carbon fibers were further characterized for mechanical properties and morphology.

501555 Sustainable Graphenic Fixed-Bed Hybrid Adsorbents: Green Synthesis and Application

Abstract:

Organic dyes are one of the most common types of pollutants found in urban waterbodies, with potentially serious health hazards. While various technologies including adsorption, precipitation, separation, amalgamation, and ion-exchange are employed to remove such contaminants from water, adsorption proved to be the most economical and efficient. Due to the high surface area and benign nature of activated carbon (AC), it is heavily used in the removal of various pollutants. However, increasing amounts of anthropogenic wastes necessitates the discovery of adsorbents with higher remediation efficacy. Graphene, the newest member of carbon-based nanomaterials, composed of sp2 hybridized carbons arranged in a perfect planar honeycomb lattice, have the highest surface area among the carbon family. Recent research indicated the large utility of graphenic materials in environmental remediation applications. However, the dependence on petroleum-based or high-value precursors for graphene synthesis is one of the challenges in its large-scale utilization for applications such as water purification. In addition, this dependence also limits the sustainability of such adsorbents. Hence, we report, *for the first time*, an immobilized graphenic adsorbent system prepared from a highly sustainable, naturally abundant, and underutilized bio-product, Lignin. Lignin, the most abundant natural polymer with aromatic monomers, was coated onto inexpensive sand particles. The lignin coated sand was thermochemically converted to graphene coated sand (GCS), leveraging the inherent high density of aromatic rings in lignin. The as-synthesized GCS with a large number of defects or void spaces, that act as the active adsorption sites, were investigated for their utility in removing a variety of dyes from simulated contaminated water. By controlling the process parameters, we engineered the microstructure of the adsorbent to derive highest possible adsorption capacity. To further embellish the adsorption efficacy, a metal-graphene hybrid adsorbent was prepared via a green reduction process. The hybrid system, due to enhanced surface area and activity, demonstrated improved adsorption capacity. It is expected that the completely green process employed for the synthesis and the highly sustainable nature of the product will attract widespread attention and help the push towards the application of graphenic materials for largescale applications such as water purification.

502694 Large-Scale Implementation of a New Technique for Sugar Extraction and Conversion to Furans from Biomass Hydrolysates

Abstract:

Lignocellulosic biomass is a sustainable feedstock for production of furans – 5-hydroxymethy furfural and furfural – versatile platform molecules that can be converted to drop-in fuels and precursors for synthetic materials. Lack of efficient pathways from lignocellulosic sugars constitutes a major barrier to the industrial production of furans. We have recently achieved high yields of furans from biomass hydrolysates using a novel hybrid enzyme- and chemo-catalytic technology. The presentation will provide proof-of-concept (bench-scale) experimental data on this novel technology as well as results from larger-scale implementation in process-modules, using biomass hydrolysates generated via AFEX (Ammonia Fiber Expansion) pretreatment.

496167 Microwave Assisted Depolymerization of Lignin to Phenolics in Polar Solvents

Abstract:

Lignin, a natural polymer is the third major constituent of vascular plants and provides structural support. It is separated from cellulose and hemicelluloses during the Kraft process in paper industry. An estimated 50 million tonnes of lignin is produced annually worldwide which is burned to generate process heat within the pulping industry [1]. The structure and aromatic units present in lignin makes it a valuable source of wide range of products ranging from macromolecules to phenolics. The conventional methods of lignin conversion require higher temperature and pressure conditions. Non-conventional methods such as microwave energy provide alternate ways of lignin conversion which are less energy intensive. Microwaves are electromagnetic radiation in the frequency range of 300MHz to 300 GHz. The interaction of microwaves with the reaction media is based on the mechanism of dipole rotation and ionic conduction. In this study, microwave assisted conversion of lignin was carried out in solvents, with high to medium polarity such as ethylene glycol (EG), dimethyl sulfoxide (DMSO) and dimethyl formamide (DMF). These solvents also have the ability to dissolve alkali lignin completely at 100°C. Experiments were carried out in a microwave reactor at different temperatures (100-140°C) and reaction time (0-80min). The effect of microwave treatment on the molecular weight distribution (MWD) and average molecular weights of lignin were analyzed using gel permeation chromatograph. The phenolics produced by lignin conversion were identified using MALDI-TOF mass spectrometry and high pressure liquid chromatography. In high polarity solvent such as EG, maximum reduction in weight average molecular weight (M_w) was observed at 120 °C, while in medium polarity solvents such as DMSO and DMF, M_w decreased with increase in temperature. The highest yield of lignin oligomers (127.74 mg $g^{-1}{}_{lignin}$) and phenolic monomers (133.7 mg $g^{-1}{}_{lignin}$) containing acetosyringone, guaiacol, syringaldehyde, anisole was obtained at 100Â°C in DMF and DMSO respectively. The possible reactions for the formation of phenolic monomers and oligomers occur by the cleavage of etheric Cβ-O and Cα-Cβ bond induced by solvent interactions. The molecular weight distribution of lignin shifted to larger molecular weights at higher temperatures and reaction time indicating the involvement of recombination reactions.

References

1. Tolbert, A.; Akinosho, H.; Khunsupat, R.; Naskar, A.K.; Ragauskas, A.J. Characterization and analysis of the molecular weight of lignin for biorefining studies. *Biofuels Bioprod. Bioref.* **2014**, *8*, 836â€"856.

494252 Spray Coating Chitin & Cellulose Nanomaterials for Enhancement of Barrier Properties

Abstract:

Chitin is an abundantly occurring biopolymer like cellulose. The annual natural chitin production is estimated to be around 10^{10} - 10^{12} tons per year (Barikani et al., 2014) and is about the same order of magnitude as cellulose, whose annual bio-production by photosynthesis is estimated to be of the order of 10^{11}-10^{12} tons per year (Klemm et al.,2004). Similar to cellulose, chitin can also occur as chitin nanofibers (ChNFs) and when it is subjected to high shear treatment, nanofibers of 10-20 nm diameter and several microns in length are obtained (Wu et al., 2014; Ifuku & Saimoto, 2012).Â Wu et al. (2014) were able to produce chitin nanofibers in suspension from purified crab Î±-chitin, and when these suspensions were cast into films the resultant films had O_2 & CO_2 gas permeabilities of 0.006 & 0.018 barrer respectively. These good barrier properties as well as chitinâ€™s natural bacterial resistance make chitin an excellent candidate for sustainable barrier packaging applications. But packaging materials are generally multilayered, and a good method for incorporating these chitin films in continuous multilayer packaging processing needs to be developed. Spray coating is one of the most versatile methods of applying liquid onto surfaces. Due to its frequent use in painting, spray coating processes are widely encountered in everyday life (McKeen, 2016). One of the main features of a spray coating process is the contactless delivery of the material to be coated onto the surface. This implies that the actual delivery process is not influenced by the surface topography and thus uneven, or delicate surfaces can be coated. The delivered load of coated material is independent of the surface and the lack of a metering gap leads to decreased sensitivity to defects and especially helps in reduction of web breaks in roll to roll process (Czerwonatis, 2008). Beneventi, et al. (2014) were able to coat slurries of micro fibrillated cellulose (MFC) onto wet highly porous papers allowing complete retention of MFC on the surface of the substrate and improved barrier properties of the final product. In the present study we demonstrate the successful spray coating of chitin nanofiber & cellulose nanocrystal (CNC) suspensions onto polylactic acid films. The spray coating of these materials was found to lead to a reduction in the oxygen barrier properties of the final composite film by as much as 75 % of its original value depending on the loading and the material used (CNC, Chitin or both). Unlike the study by Beneventi, et al. the substrate in this case is non porous and the suspensions do not wet the surface. The resulting de-wetting problems were successfully solved by heating the substrate surface. This process could open up a path for the use of more sustainable polymers with otherwise excellent properties but are lacking in barrier properties. Addition of a ChNF or CNC layer could also lead to a reduction in the amount of the more expensive polymer in the multilayer (with the sole function of providing oxygen barrier) all the while increasing the sustainability.

500752 Development of Low-Concentration Alkaline Treatment Method to Produce Thermally Stable Cellulose Nanofibrils (CNFs)

Abstract:

The thermal stability of cellulose nanofibrils (CNFs) can be improved by converting cellulose crystal structure from cellulose I to cellulose II using alkaline pretreatment method. The conventional pretreatment method requires 17.5% or higher concentrations of sodium hydroxide and large volume of water for neutralization to produce thermally stable, but interdigitated or aggregated CNFs. The main objective of this study was to develop a low concentration alkaline pretreatment method to produce thermally stable and highly dispersible CNFs. Cellulose nanofibrils with improved thermal stability were successfully prepared by treating cellulose powder with 2wt.% NaOH solution at low temperature (below 0), followed by mechanical disintegration through a French cell press with only ten passes. The thermally stable CNFs had an average diameter of 90 nm with the mean onset and maximum thermal decomposition temperature of 305 and 343 respectively, compared with those of 283 and 310 respectively, for the CNFs prepared without the NaOH treatment. The thermally stable CNFs can create new opportunities for the development of CNFs based packaging composites and high-end electronic applications.

491943 Tailored and Integrated Production of Carboxylated and Lignin Containing Cellulose Nanocrystals and Nanofibrils for Composite Applications

Abstract:

Here we demonstrate di-carboxylic acid hydrolysis for the integrated production of lignin containing cellulose nanocrystals (LCNC) and nanofibrils (LCNF) using two unbleached mixed hardwood chemical pulps of lignin contents of 3.9 and 17.2%. Acid hydrolysis experiments used maleic acid solution of 60 wt% concentration at 120°C for 120 min under ambient pressure. Yields of LCNC were low of less than 6% under this set of condition. The higher lignin content samples produced LCNC with greater height (diameter) of 25 nm but similar length of approximately 230 nm to that from the lower lignin content fibers (height of 20 nm). Interestingly, the higher lignin content sample resulted in LCNF with smaller height (diameter) of 7 nm but longer length of > 1 μm, or greater aspect ratio than the LCNF from the lower lignin fibers of height 10 nm and length < 1 μm. Lignin protected cellulose from esterification which resulted in LCNC and LCNF that was less carboxylated compared to those lignin-free CNC and CNF and therefore had lower charges. However, lignin is more hydrophobic and thermally stable than carbohydrates therefore LCNC and LCNF are favorable for composite applications.

504076 Nano Carbon Structures from Cellulosic Biomass for Use As Functional Materials Â

Abstract:

We report herein the production of hierarchical nanocarbon materials from cellulose fibers. The carbonized cellulosic fibers with branching out CNTs (CC-CNT) offer interesting surface and electrochemical properties that are of interests for chemical processing and biosensing applications. In our tests for chemical adsorption tests, CC-CNT offered highly selective adsorption of hydrocarbons from gas mixtures, especially chemicals with C=C bond and aromatic structures. More interestingly, the chemicals absorbed and accumulated can be quickly released from the surface of the materials upon application of modest electrical current. Such an electrically switchable adsorption behavior may promise applications in a broad areas including air quality control, chemical sensing and fuel storage.

Rowan University

Novel Bio-Based Polyesters & Polycarbonates from Xylochemicals

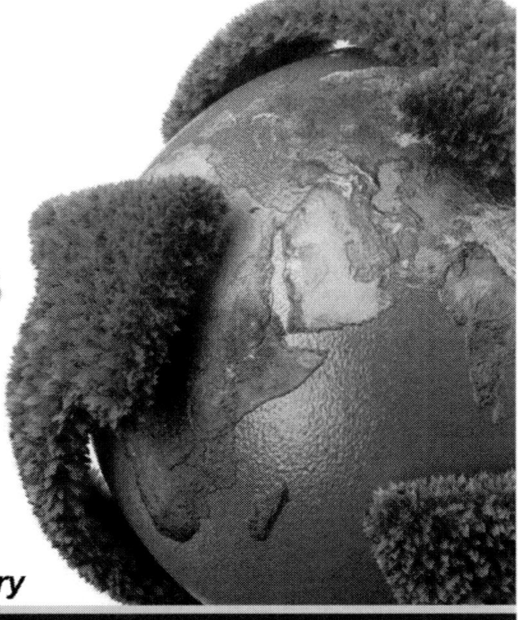

2017 AIChE Annual Meeting

Minneapolis, MN October 29 – November 3 2017

"Bioplastics, Biocomposites and Value-Added Uses of Biofuel Coproducts for Sustainable Manufacturing"

Joseph F. Stanzione, III, Ph.D.

Department of Chemical Engineering

Sustainable Materials Research Laboratory

 1

Plastics: The Facts

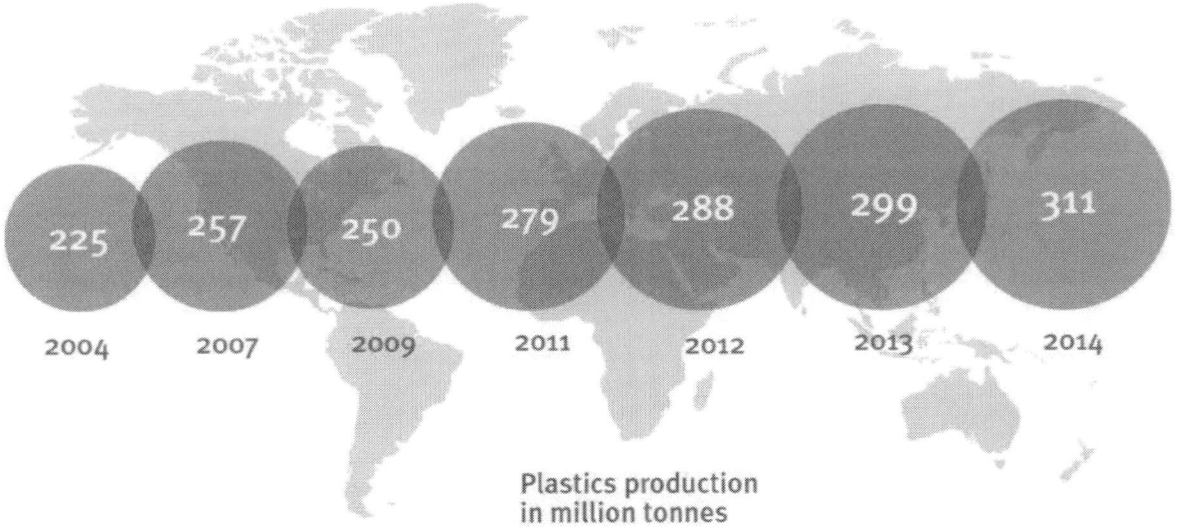

- **Plastic production is growing globally[1]**
- **Currently around 330 Mtonnes**

[1] PlasticsEurope, *Plastics - the Facts* (2015)

 2

Plastics: The Facts

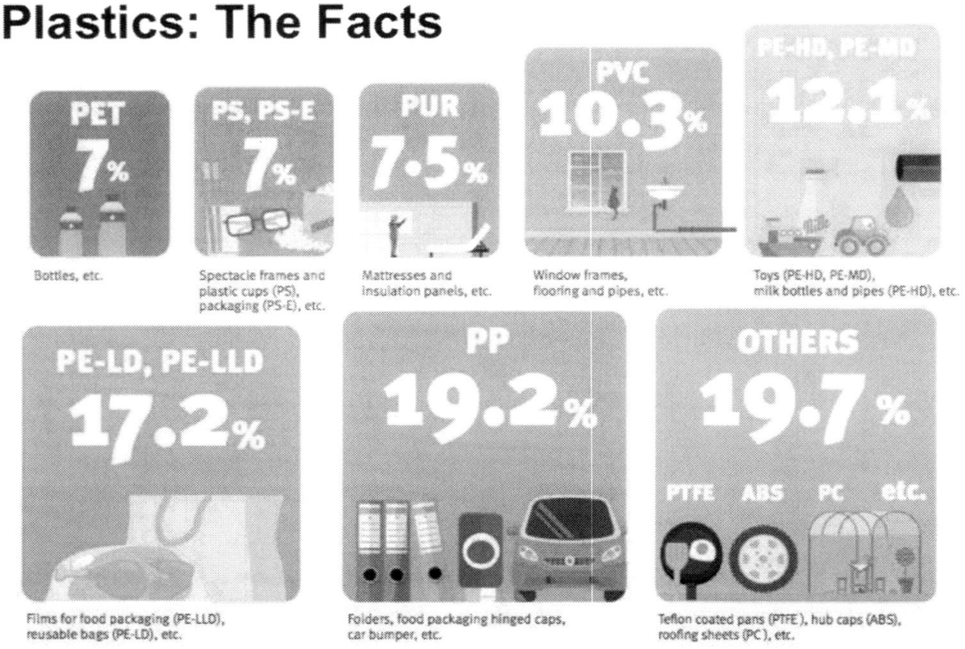

Less than 1.5% are renewably sourced[2]

[2] European Bioplastics, *Bioplastics facts and figures* (2017)

 3

Plastics: The Future

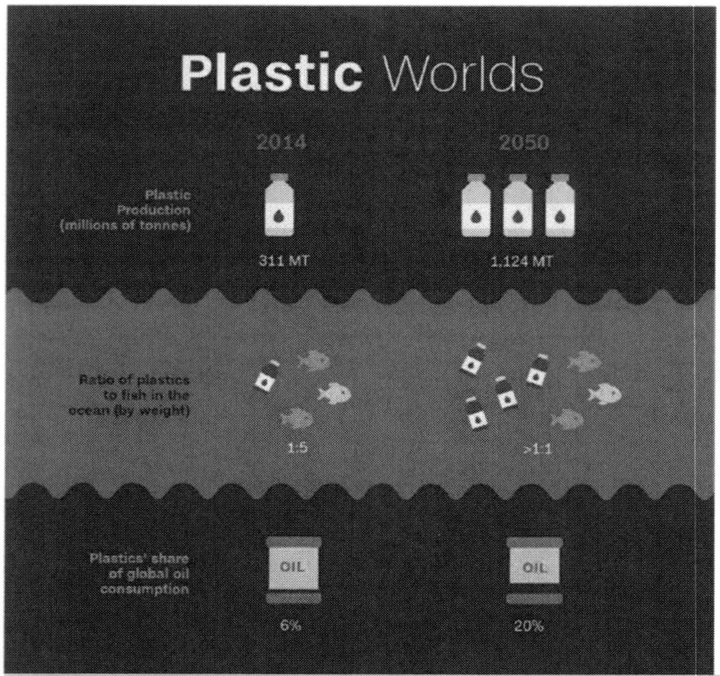

By 2050:

- **Expected to triple**
- **More plastic than fish in the ocean**
- **20% of global annual oil consumption**

[3] World Economic Forum (2016)

 4

Trying to Replace PET

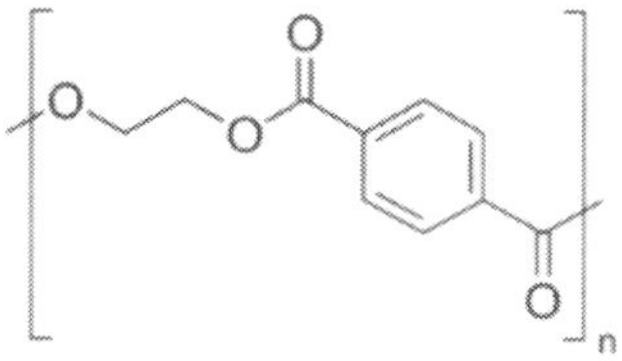

- **Very good mechanical properties**
- **T_g = (67-81) °C, T_m ≈ 267 °C**
- Petroleum-derived (green routes in progress)
- Not easily degraded

High Tg Bio-Based Aliphatic Polyesters from Bicyclic D-Mannitol

Cristina Lavilla, Abdelilah Alla, Antxon Martínez de Ilarduya, and Sebastián Muñoz-Guerra*

Departament d'Enginyeria Química, Universitat Politècnica de Catalunya, ETSEIB, Diagonal 647, 08028 Barcelona, Spain

- Sugar derived
- T_gs range from -40 to 70 °C

Renewable (semi)aromatic polyesters from symmetrical vanillin-based dimers†

Audrey Llevot,[a,b] Etienne Grau,[a,b] Stéphane Carlotti,[a,b] Stéphane Grelier[a,b] and Henri Cramail*[a,b]

- Lignin derived
- T_gs up to 140 °C
- Insoluble in any solvent

Bio-based alternatives in the synthesis of aliphatic–aromatic polyesters dedicated to biodegradable film applications

Nicolas Jacquel [c], René Saint-Loup [c], Jean-Pierre Pascault [a, b], Alain Rousseau [a, b], Françoise Fenouillot [a, b, *]

[a] Université de Lyon, F-69003, France
[b] INSA-Lyon, Ingénierie des Matériaux Polymères, IMP, UMR5223, F-69621 Villeurbanne, France
[c] Roquette Frères, Division Recherche, F-62080 Lestrem, France

- Sugar derived (isosorbide)
- Negative T_gs, high T_ms

etc...!

 5

Additional Concerns: Toxicity

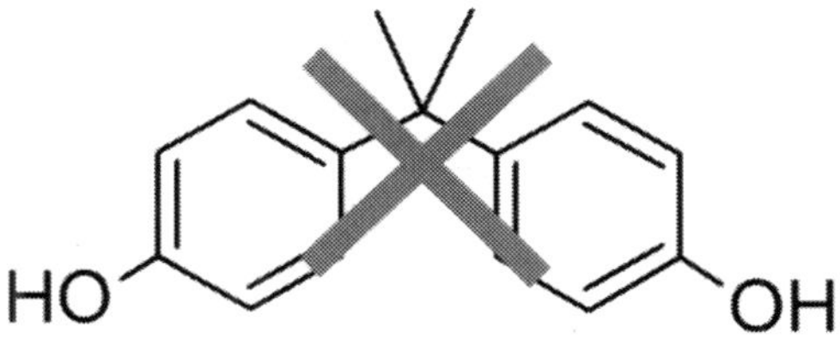

Bisphenol A (BPA)

 6

Bisguaiacol (BG): A Lignin-Derived Bisphenol

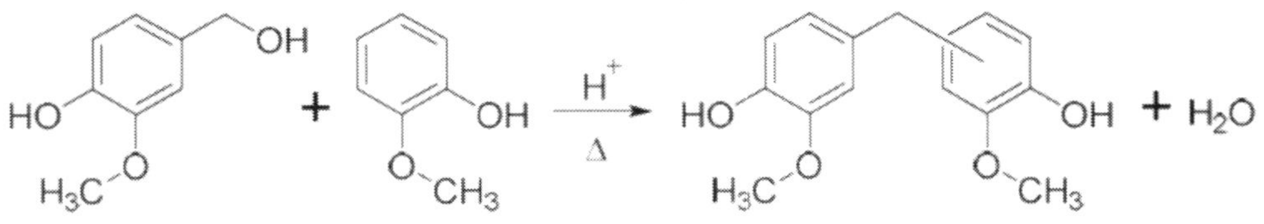

- Similar structure to BPA[4]
- Fully renewable
- Reduced toxicity[5]
- Isomeric distribution
  - 82% *p,p'*
  - 15% *m,p'*
  - 3% *o,p'*

[4] Hernandez *et al.*, *ACS Sustain Chem Eng* (2016)
[5] Hong *et al.*, *Int. J. Environ. Res. Publ. Health*(2016)

Q: Can we use BG to synthesize high-performance renewable thermoplastics?

BG-based Polycarbonates

Joseph Mauck,
M.S. CHE

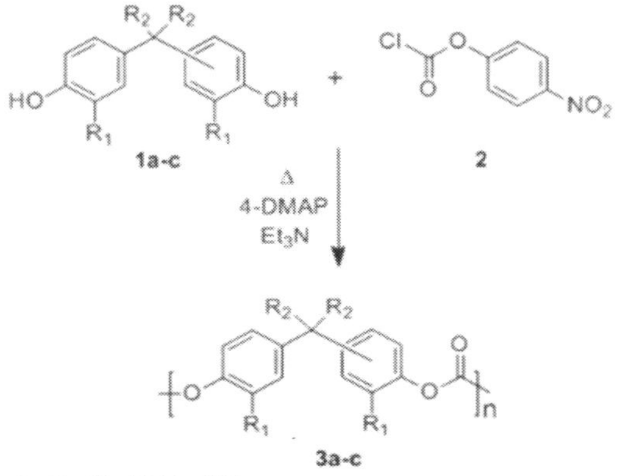

1a,3a: R_1=H, R_2=CH_3
1b,3b: R_1=H, R_2=H
1c,3c: R_1=OCH_3, R_2=H

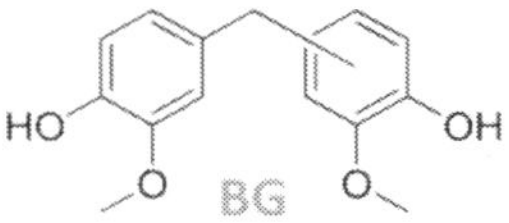

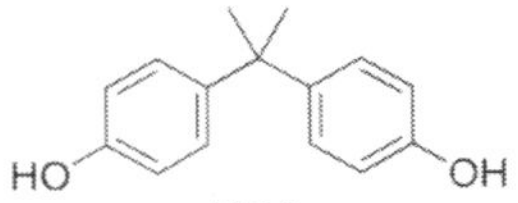

BPA

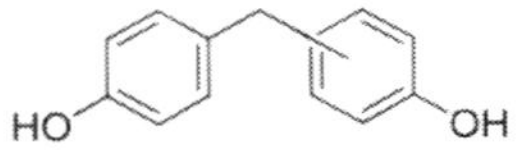

BPF

[6] J. Mauck *et al.*, *J Biobased Materials & Bioenergy* (2017)

BG-based Polycarbonates: Thermal Properties

Sample	*IDT* (°C)	$T_{50\%}$ (°C)	T_{max} (°C)	Char Content (%)	T_g (°C)
BG-PC	375	455	450	24.2	100
BPA-PC	440	543	545	23.2	150
BPF-PC	369	546	501	40.5	119

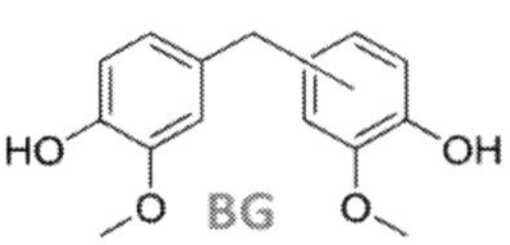

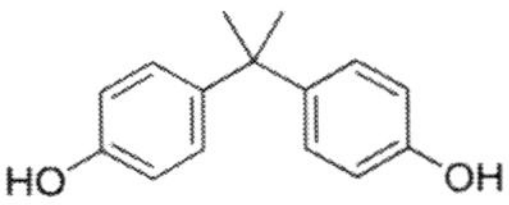

BPA

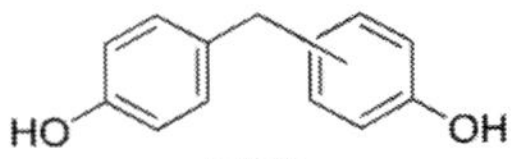

BPF

[6] J. Mauck *et al.*, *J Biobased Materials & Bioenergy* (2017)

Q: Can we use BG to synthesize polyesters?

Preparation of BG-based Polyesters

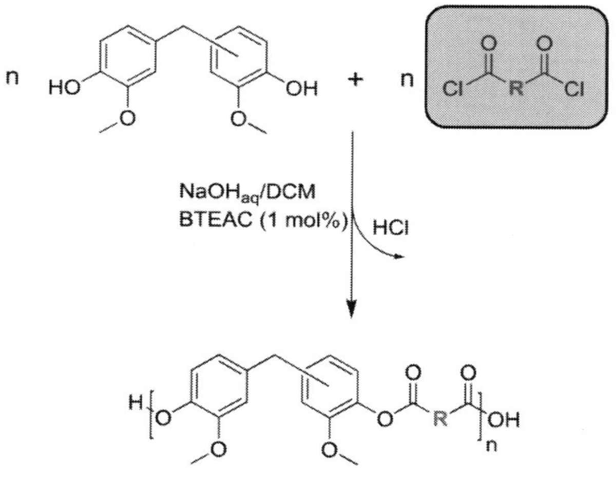

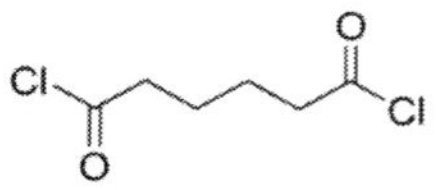

Adipoyl Chloride

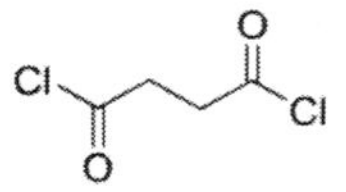

Succinyl Chloride

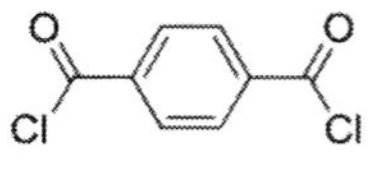

Terephthaloyl Chloride

[7] S. Curia *et al.*, *In preparation* (2017)

BG-based polyesters

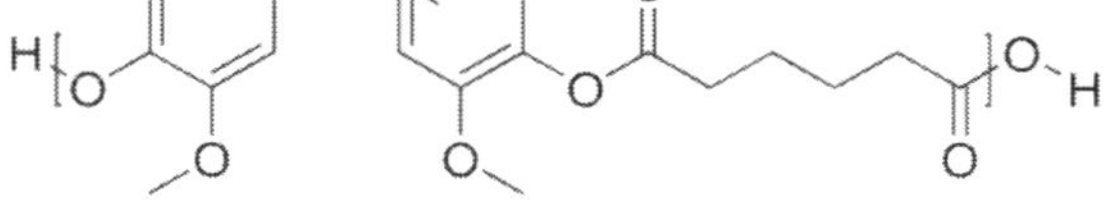

Poly(BG adip)
M_n=6.6 kg mol^{-1} (Đ=1.82)

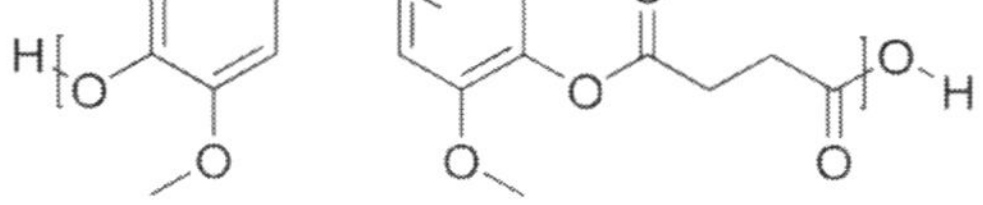

Poly(BG succ)
M_n=7.2 kg mol^{-1} (Đ=2.16)

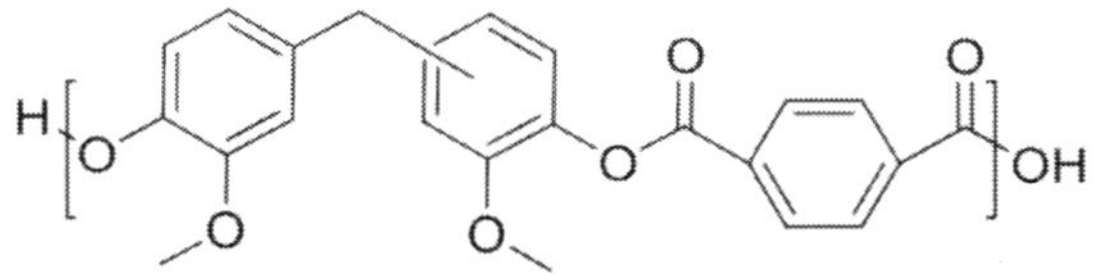

Poly(BG tere)
M_n=26.9 kg mol^{-1} (Đ=2.29)

[7] S. Curia *et al.*, *In preparation* (2017)

IR Analysis

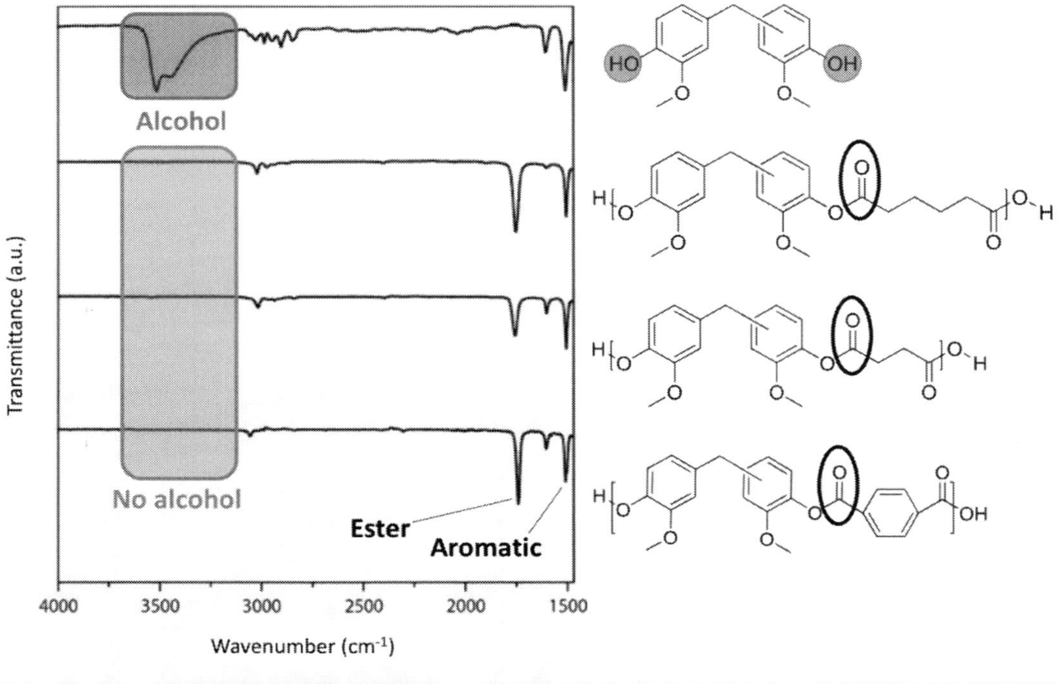

^{1}H-NMR Analysis (in $CDCl_3$)

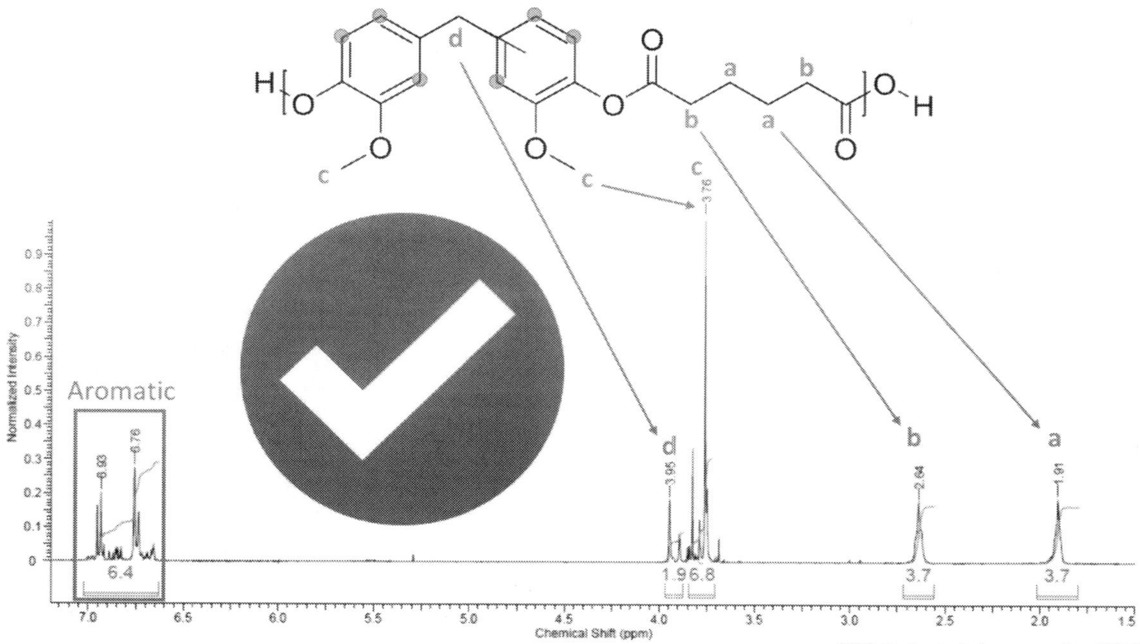

[7] S. Curia *et al.*, *In preparation* (2017)

BG-based Polyesters: DSC Analysis

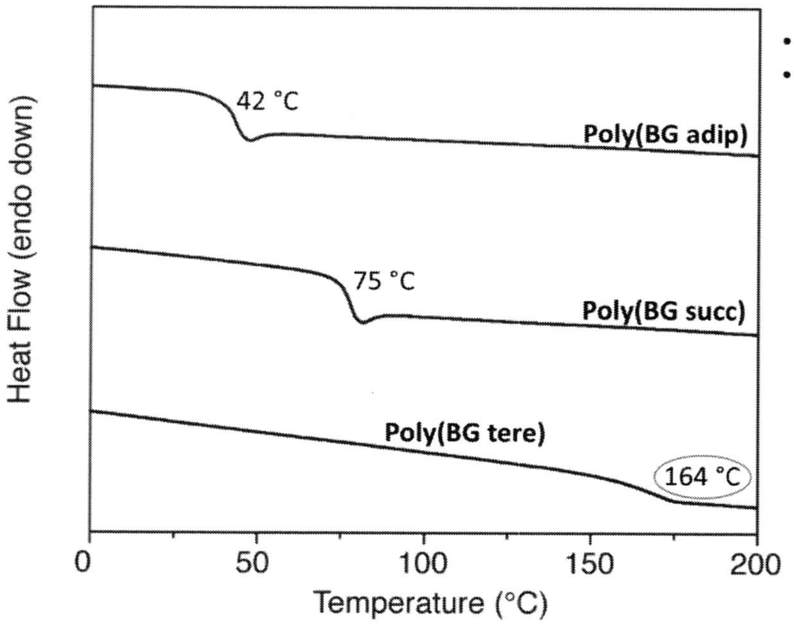

- **10 °C min^{-1}**
- **N_2 50 mL min^{-1}**

[7] S. Curia *et al.*, *In preparation* (2017)

BG-based Polyesters: Thermal Stability (N_2)

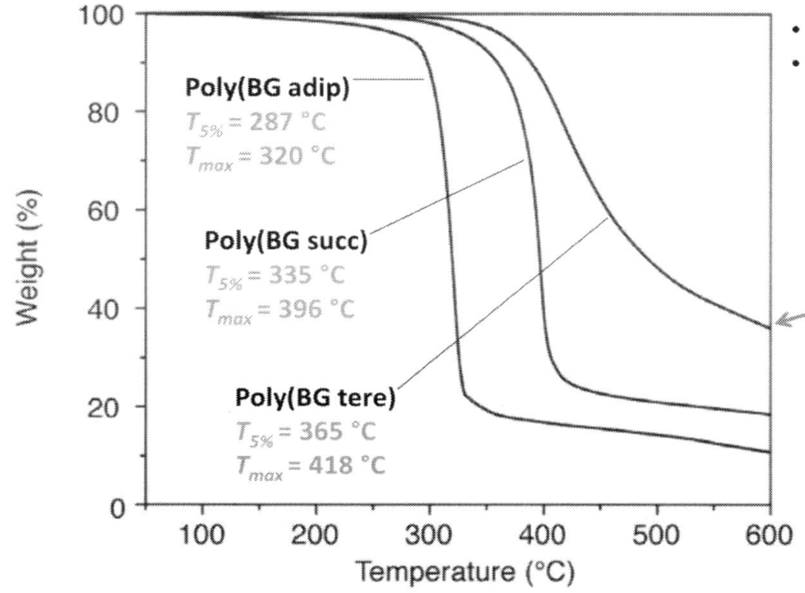

- 10 °C min^{-1}
- N_2 25 mL min^{-1}

[7] S. Curia *et al.*, *In preparation* (2017)

BG-based Polyesters: Thermal Stability (Air)

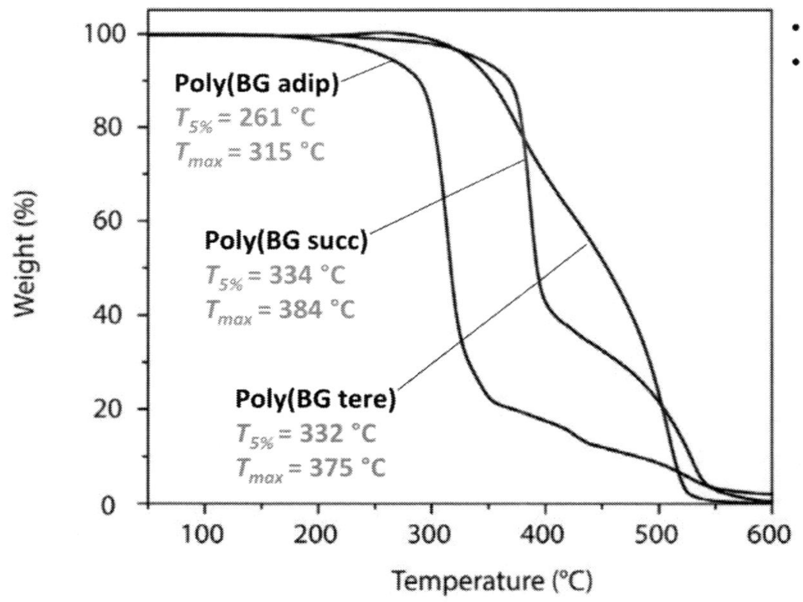

- 10 °C min^{-1}
- Air 25 mL min^{-1}

T_{max} still above 300 °C

[7] S. Curia *et al.*, *In preparation* (2017)

BG+FDCA: a Fully Renewable Polyarylate

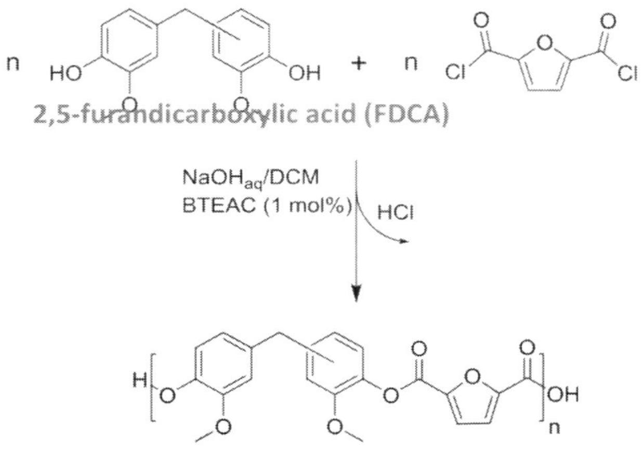

Poly(BG FDCA)
M_n=6.3 kg mol^{-1} (Đ=1.57)

[7] S. Curia *et al.*, *In preparation* (2017)

Poly(BG-FDCA): DSC

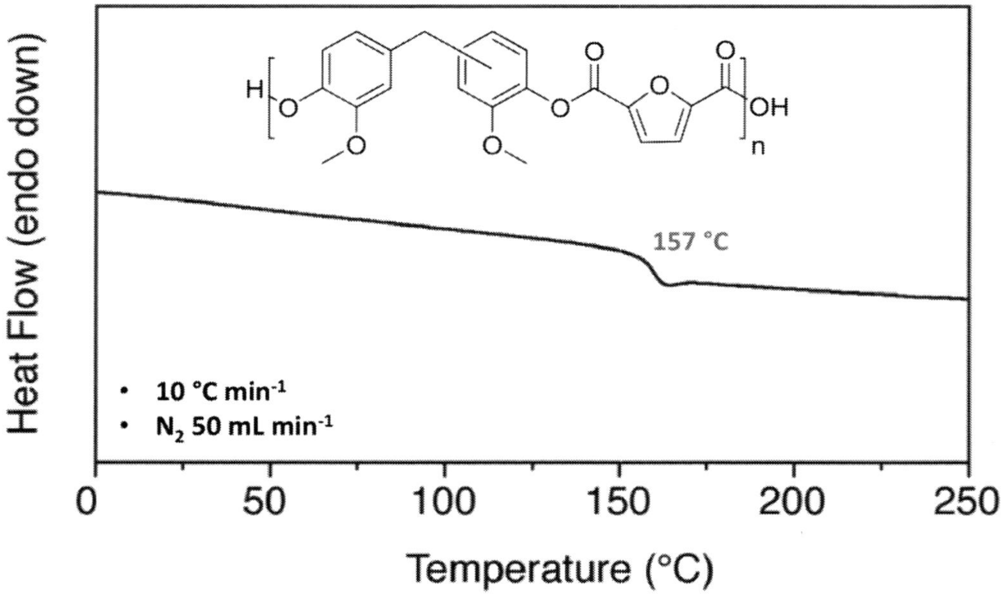

[7] S. Curia *et al.*, *In preparation* (2017)

Poly(BG-FDCA): TGA N_2 *vs* air

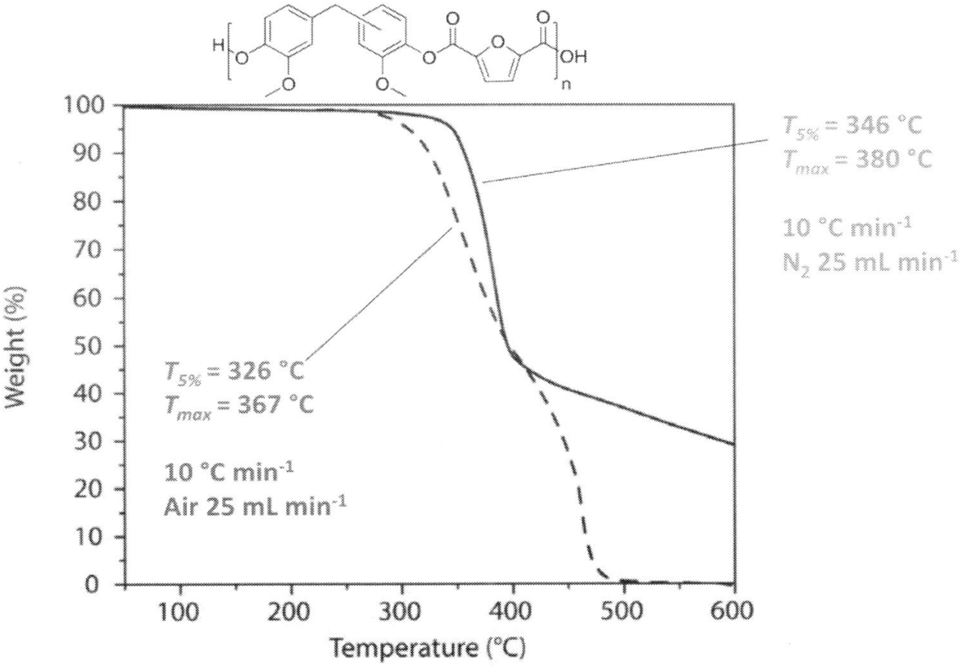

[7] S. Curia *et al.*, *In preparation* (2017)

Conclusions

- **BG** can be used to synthesise **useful polycarbonates** & **polyesters**
- **BG-based polyesters** show **high thermal resistance** and **tuneable T_g**
- **Poly(BG FDCA)** has a remarkably **high T_g (157 °C)**

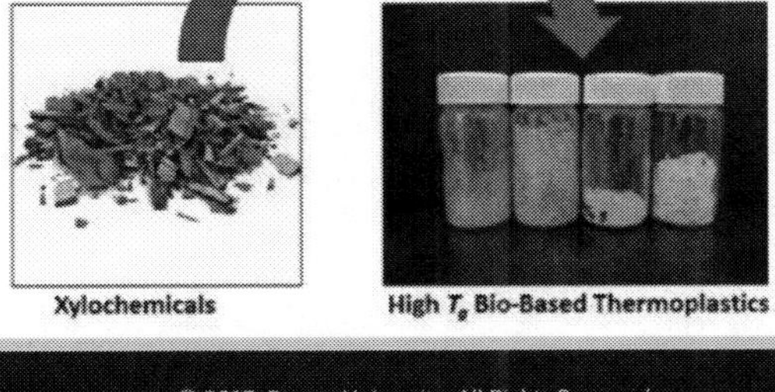

Future Investigations

- Hydrolytic degradation & biodegradation tests
- End-cap with functional molecules
- Chain extension, melt polymerization (achieve higher MW)
- Gas permeability/diffusivity
- Mechanical tests

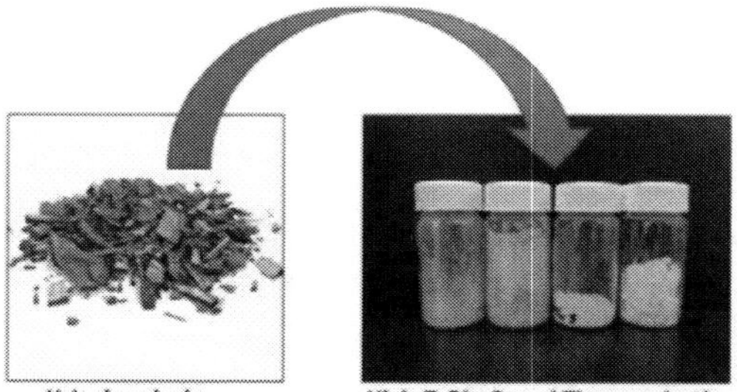

 23

Acknowledgements

- Sustainable Materials Research Laboratory
- Joseph R. Mauck
- Silvio Curia, Ph.D.
- John J. La Scala, Ph.D. (ARL)
- Rowan University
  - ❖ Department of Chemical Engineering
  - ❖ Department of Chemistry & Biochemistry
- U.S. ARL for Financial Support via Cooperative Agreement W911NF-16-2-0225

Silvio Curia, Ph.D.

Rowan University

ARL U.S. ARMY RDECOM

South Jersey Technology Park at Rowan University

 24

Thank you!
Any questions?

Crater Lake National Park

Rocky Mountain National Park

503981 Light Weight Biocomposite from Toughened Polyolefin and Biocarbon

Abstract:

In converting biomass to bioenergy through thermochemical processes, a considerable amount of solid residue generates in each conversion cycle (up to 35 wt.%) which mainly consists of renewable carbon (biocarbon). However, the biocarbon is considered as a byproduct of the pyrolysis units. Unless a high-volume application exists for the biocarbon, the significant amount of this byproduct can challenge the long-term sustainability of pyrolysis units. In this project, biocarbon is intended to be used as filler in plastic industry targeting toughened polypropylene (PP) based automotive parts. In order to achieve a suitable balance of mechanical properties, biocomposites were compatibilized using a maleated polypropylene. Furthermore, the PP phase was nucleated using a Î² nucleating agent to increase the impact toughness without significant loss in the stiffness. Tensile properties and notched Izod impact strength were measured as response variable. Atomic force microscopy (AFM) was utilized to analyze the PP-biocarbon interface in presence and absence of the compatibilizer. The effect of the nucleating agent was investigated by differential scanning calorimetry (DSC) and correlated to the improvement of biocomposites impact toughness. The results suggest that the morphological changes caused by the compatibilizer and toughening effect of Î² crystals can generate high impact toughness in composites with 20 wt.% loaded biocarbon. Therefore, the biocarbon can be utilized in high loading levels similar to current petroleum-based composites with comparable mechanical properties. At the end, substituting conventional mineral filled composites with biocarbon-based biocomposites can provide the required mechanical properties with added benefits of lower density (â‰ˆ15%), 30% higher bio-based content and cheaper price. These benefits would translate into improved fuel economy of cars and reduction in greenhouse gas production. Considering the high-volume usage of plastic parts especially in automotive industry, this approach would make the biomass to bioenergy conversion more sustainable.

Keywords: Biocarbon, Toughened biocomposites, Polypropylene.

Acknowledgement: We gratefully acknowledge the financial support of the Ontario Ministry of Agriculture, Food and Rural Affairs (OMAFRA)/University of Guelph - Bioeconomy for Industrial Uses Research Program Theme Project # 030055; Ontario Research Fund, Research Excellence Program; Round-7 (ORF-RE07) from the Ontario Ministry of Research, Innovation and Science (MRIS) (Project #052644 and 052665), and the Natural Sciences and Engineering Research Council (NSERC), Canada Discovery Grant Project # 401111 to carry out this research work

497443 Secondary Fermentation of Corn Ethanol Co-Products for Improved Amino Acid Qualities

Abstract:

Secondary Fermentation of Corn Ethanol Co-Products for improved Amino Acid Qualities

In 2016 5.28 billion bushels of corn were used to produce about 14.79 billion gallons of ethanol in the United States. As a result, about 36 million tons of Dried Distillers Grains with Solubles (DDGS) were manufactured and fed to livestock (U.S. Dept. of Energy, 2017). DDGS are a common feed supplement in cattle rations as it is inexpensive and has positive feeding characteristics. One of the drawbacks of DDGS is that it is lacking in in key amino acids such as tryptophan, arginine, and lysine. Historically these amino acids have been supplemented by external addition of feed grade amino acids to rations. The research carried out attempts to fortify DDGS with higher amounts of tryptophan through secondary fermentation of Whole Stillage (WS) by bacteria. When cultured on WS the bacteria secretes extracellular tryptophan, which serves to close the amino acid gap in corn ethanol co-products. The bacteria used is Generally Regarded as Safe (GRAS) and has been used to produce feed ingredients historically. Because of the research, profitability of corn ethanol operations will be increased due to the improved feeding value of the co-products produced.

References

United State Department of Energy, â€œEthanol Fuel Basics,â€ Energy Efficiency & Renewable Energy http://www.afdc.energy.gov/fuels/ethanol_fuel_basics.html 03/16/2017

500583 Extraction High-Value Chemicals from Ethanol Co-Products: A Feasibility Assay on Phytate Extraction with Life Cycle and Techno Economic Assessment

Abstract:

Background

The well-established dry-grind process today accounts for also over 80% of the operating ethanol facilities in the United States, producing not only ethanol, but also distiller's grains. Distiller's grains, commonly commercialized as dried distiller's grains with solubles (DDGS) accounts for most of the heavy fraction from the ethanol distillation and rectification operations, and is processed through a complex downstream processing after the fermentation is unloaded. Despite being a nutritious animal feed, inclusion of DDGS as feed ration is still limited due to compositional properties, such as the high presence of non-starch polysaccharides, which are non-digestible to monogastric animals, the fatty acid profile, and the overdose of phosphorus, which is largely non-available to some animals. Phosphorus behavior in ethanol downstream processing is of particular interest, not only because it can lead to environmental issues, but also because it can potentially be an additional economic source to ethanol producers. Over 40% of the phosphorus present in ethanol co-products is present as phytate, the main form of organic phosphorus in cereals and grains, which is not completely hydrolyzed during the whole ethanol process. Phytate has a high market value, and today is a platform chemical utilized in a wide array of industries, and it is mostly exploited because of its high chelating potential (He et al. 2017). Phytate is also the raw material to produce *myo*-inositol, a cyclic polyol with a variety of uses by pharmaceutical and nutraceutical companies.

Phytate is found to be dissolved on the ethanol co-products(He et al. 2017). During the dry-grind downstream processing, thin stillage, i.e., the liquid-rich fraction derived from the centrifugation of whole stillage, accounts for the highest concentration of phytate among all the co-products. Recent research (He et al. 2017) has proposed an extraction method using ion exchange separation, which can efficiently separate phytate from thin stillage, with little to no effect on the residual nutritional profile of the final DDGS. Even though the concentration of phytate being relatively low on ethanol co-products, at around 1 g L^{-1} in thin stillage, due to the enormous volumetric production by dry-grind plants, the production leveraging costs in extracting phytate might become an industrial alternative.

The high volumes of DDGS produced by ethanol plants, alongside with its limited dietary inclusion, pose a challenge for higher revenues to ethanol plants. The second-most important co-product from dry-grind ethanol facilities, distiller's corn oil, represent today a mature product suitable for biodiesel production and other uses of low-grade vegetable oil(Reis et al. 2017). Removal of distiller's corn oil, on the other hand, has been criticized by animal nutritionists claiming a loss in energy value from DDGS. Despite still representing about 20% of the total revenue stream from dry-grind ethanol plants, DDGS are still ranking in low positions when compared to higher end animal feeds, such as soybean meal. The extraction of high-value chemicals through mature sound technologies may represent, thus, an alternative for higher economic security of ethanol plants. This current research describes a techno-economic analysis and a life-cycle assessment of a phytate extraction process on a large-scale dry-grind corn-to-ethanol plant in the United States, and its posterior *in-loco* conversion to *myo*-inositol.

Methods

The scope of this work analyzed the utilization of corn-to-ethanol thin stillage produced through the dry-grind process as raw material to extract phytate using an ion exchange process, followed by hydrolysis to generate *myo*-inositol. The process was modelled according to He et al. (2017) , and by additional empirical analysis, when required. Thin stillage was obtained from a dry-grind ethanol plant in the state of Iowa, USA. The process of converting thin stillage to a phytate brine solution is divided into the major steps: (1) filtration, (2) flow through ion exchange column, (3) desorption of phytate using brine solution, (4) reconditioning of column using sodium hydroxide, and (5) column washing using water. The conversion of phytate brine solution to calcium phytate is done by (6) addition of calcium chloride and sodium hydroxide, followed by product filtration, and production of *myo*-inositol is achieved by (7) hydrolysis of phytate brine solution, (8) precipitation of phosphate with calcium chloride and sodium hydroxide, and (9) evaporation and drying of *myo*-inositol.

The life cycle assessment of calcium phytate extracted from corn-to-ethanol thin stillage was estimated using GREET model. Calcium phytate extraction, and posterior conversion to *myo*-inositol were implemented on two stationary models on GREET – Dry mill ethanol production w/ corn oil extraction, and Dry mill ethanol production w/o corn oil extraction. Techno-economic assessment for ethanol plants was considered using the BuGal model, with added market and technical data for the modified phytate extraction and conversion units. Since the basis of the BuGal model for techno-economic analysis has been modified to analyze both conditions (with and without distiller's corn oil extraction), the two stationary models on GREET to describe corn-to-ethanol process were utilized.

Results and discussion

LCA results indicate that the proposed phytate extraction does not impact significantly most of the environmental impact factors of ethanol. As a gallon of ethanol was considered as functional unit, the results indicate that for a system with phytate extraction and oil extraction, CO_2 emissions will increase from 3.21 kg to 3.28 kg per gallon of ethanol, whilst for a system without distiller's corn oil extraction, 3.23 kg of CO_2 will be increased to 3.30 kg of CO_2 per gallon of ethanol produced. NOx, SOx, N2O, and other gases also increased with phytate extraction by a factor of about 1%. Analyses were carried out for techno economic assessment in 4 different cases, (1) with distiller's corn oil extraction and phytate extraction, (2) with distiller's corn oil extraction and without phytate extraction, (3) without oil extraction and without phytate extraction extraction, (4) without oil extraction and with phytate extraction, and were consistently all across different cases.

Unlike major recent upgrades to the dry-grind ethanol industry, such as distiller's corn oil removal, extraction of phytate and production of *myo*-inositol are related to products found in dilute concentrations. The low concentration of phytate in ethanol co-products requires low energy input on an ethanol basis, while still providing a high-value co-product. BuGal model calculates the operating expenses per gallon of denatured ethanol sold as USD 0.2673 for the control conditions, i.e., without phytate extraction, and USD 0.2686 for a system considering 20% of thin stillage being redirected to phytate extraction, due to increase in energy consumption for additional processing. Regarding chemical costs, phytate extraction and conversion to inositol likely increased the value from USD 0.1201 to USD 0.1226 per gallon of denatured ethanol, mainly due to the requirement of NaCl for phytate desorption from the ion exchange process, additional process water, and calcium chloride addition. These costs are balanced by the calcium phytate, *myo*-inositol, and calcium phosphate as fertilizer (derived from the phytate hydrolysis) sales, at prices estimated at USD 12, USD 10, and USD 0.20 per kg, respectively. BuGal estimates an average large-scale dry-grind operation, with factor of equity and debt being 0.7 and 0.3, respectively, with an interest rate charged on debt as 0.07, and the rate of return required by investors on the plant's equity as 0.12. Even though these numbers are variable, the economic considerations on such assumptions are beyond the scope of this work, and remain constant through all assumptions. Following similar numbers as Kwiatkowski et al. (2006), 2.795 gallons of ethanol are estimated to be produced from each bushel of corn, and 17.5 lb of

DDGS, and CO_2 are produced on a bushel basis. Denatured ethanol sales are estimated to be at USD 1.70 per gallon, an estimate within current market prices , and that phytate extraction would not affect the price of DDGS, thus, being marketed at a price of USD 161 per ton, also within current market estimates.

Conclusion

The removal of phytate from thin stillage may represent a novel process in the dry-grind ethanol industry. The removal of phytate via ion exchange does not increase significantly (around 1% of increase) the emissions involved on the production of one gallon of ethanol, and that it can be a source of profit if operated within a certain range of thin stillage processed within a plant. The results also presented herein describe the potential of converting the extracted phytate to myo-inositol. Myo-inositol has a more flexible market when compared to phytate, and could also represent another source of revenue for ethanol plants.

References

Kwiatkowski JR, McAloon AJ, Taylor F, Johnston DB (2006) Modeling the process and costs of fuel ethanol production by the corn dry-grind process Industrial crops and products 23:288-296

He Q, Rodrigues Reis, CE, Wang F, Hu B (2017) Phytate extraction from coproducts of the dry-grind corn ethanol process. RSC Advances. doi:10.1039/C6RA27409A

Reis CER, Rajendran A, Hu B (2017) New technologies in value addition to the thin stillage from corn-to-ethanol process Reviews in Environmental Science and Bio/Technology 16:175-206 doi:10.1007/s11157-017-9421-6

502221 Selective Lignin, Cellulose, and Hemicellulose Dissolution in Deep Eutectic Solvents

Abstract:

An method to separate cellulose and hemicelluloses from lignin in biomass that is environmentally-friendly would be helpful in reducing greenhouse gas generation. Deep eutectic solvents (DES) are easily prepared and have low volatility, non-flammability, biocompatibility, and biodegradability. Some DESs were researched to determine their capability to preferentially dissolve lignin at 60 . Thermogravimetric analysis of these DES showed them to be stable at typical biomass processing temperatures. Pretreating loblolly pine in one DES enhanced glucose yield after enzymatic hydrolysis to be more than seven times that of raw or glycerol-pretreated pine. The density of DES-pretreated biomass was found to be 40% higher than the untreated pine's density.

499232 Assessing the Dispersion Influence of Cellulose Nanofibers on Papermaking Applications

Abstract:

Nanocellulose, a unique and promising natural material extracted from native cellulose, has gained much attention for its use as reinforcing agent in several matrix, because of its remarkable mechanical and physical properties, special surface chemistry and excellent environmental properties (biocompatibility, biodegradability and low toxicity). Generally, the family of nanocellulose can be divided in three types, (1) cellulose nanocrystals (CNC), with the synonyms of nanocrystalline cellulose, cellulose whiskers, rod-like cellulose microcrystals; (2) cellulose nanofibers (CNF), with other designations such as nanofibrillated cellulose or cellulose nanofibrils; and (3) bacterial cellulose (BC), which refers to nanocellulose produced by bacteria [1].

The sources for CNC and CNF extraction are wood, cotton, hemp, flax, wheat straw, sugar beet, potato tuber, mulberry bark, ramie, algae, and tunicin [2-4]. Acid hydrolysis is commonly performed for the extraction of CNC from native cellulose, through the removal of amorphous regions and preservation of highly-crystalline structure [5]. Some of the advantages that CNC suspensions are the self-assembly, very high strength due to the high crystallinity, high functionalization for its inherently high surface area, unique optical properties and lightweight [6]. Recently, CNC have become an extremely important class of nanomaterials to reinforce pulp suspensions [7-9]. Sulfuric acid hydrolysis produced stable CNC suspensions due to sulfate groups formed at the nanocrystal surface [10]. The presence of these negatively charged groups induces the formation of a negative electrostatic layer covering the nanocrystals and promotes their dispersion in water but the stability commonly declines over time, forming aggregates that sediment and are difficult to redisperse [1].

Regarding the preparation of CNF, mechanical defibrillation is applied to reduce the diameter of cellulose fibers, keeping longer nanofiber length; which involves high-pressure homogenization or microfluidization combined with chemical or enzymatic pretreatments, which reduced the energy consumption of the mechanical process. Due to the entanglement of long cellulosic chains, it is not so easy to determine the length of CNF, commonly higher than 1Â Î¼m, with microscopic techniques. CNF suspensions are distinguished from CNC by having higher aspect ratio, lower crystallinity index, higher water absorption capacity and lower light transmittance [11]. High nanofibrillation yield of CNF suspensions implies homogenization at consistencies between 1-3 wt% obtaining a stable gel structure which is difficult to disperse in separated nanofibers[12]. The highÂ viscosity of CNF suspensionÂ at low concentration makes very interesting as a non-caloric stabilizer and gellant in food applications but affect negatively when CNF is used as a reinforcement agent in a fiber network.

Several studies have evaluated the ability of CNF to improve mechanical paper properties in virgin [13-15] and recycled pulps [16-18]. Moreover, some authors reported the relationship between poor dispersion of the CNF gel into the recycled pulp and lower tensile[16], but extensive studies about nanocellulose dispersion are still needed.

Recently studies have demonstrated the feasible of flexographic ink removal using TEMPO-oxidized CNF hydrogels through its combination with cationic polyacrylamide (cPAM). The removal water-based ink mechanism is based on the adsorption of ink particles onto the CNF hydrogel followed by the flocculation of the system CNF-ink particles with cPAM addition[19]. However, the shear force generated during stirring had not been studied and could have an important impact in CNF gel structure and, therefore, in the removal ink mechanism.

Therefore, the dispersion of nanocellulose suspensions has rarely been investigated and continues to be a challenge for industrial implementation.

The objective of this study is to establish how the dispersion of the nanocellulose influence the final properties of the recycled paper and the water-based ink removal. On the one side, the effect of different pulping conditions and the use of dispersant agents were studied in presence of two retention additives. Results showed pulping conditions have an important impact on dispersion and retention of CNC and CNF, although it is dependent on the retention additive used. Moreover, the use of dispersant agent favored the dispersion of nanocellulose in the pulp, increasing tensile index of the recycled paper in 20.6% and 11.1%; using 1.5% CNF and 3.0% CNC, respectively. On the other hand, results proved that high stirring rate (750 rpm) during the interaction CNF-ink particles decreased the removal ink efficiency.

References

1. N. Lin and A. Dufresne, *Eur. Polym. J.*, 2014, **59**, 302-325.

2. I. Besbes, S. Alila and S. Boufi, *Carbohydr. Polym.*, 2011, **84**, 975-983.

3. M. Munk, H. M. Brandao, S. Nowak, L. Mouton, J. C. Gern, A. S. Guimaraes, C. Yepremian, A. Coute, N. R. B. Raposo, J. M. Marconcini and R. Brayner, *Ecotox. Environ. Safe.*, 2015, **122**, 399-405.

4. M. Jonoobi, R. Oladi, Y. Davoudpour, K. Oksman, A. Dufresne, Y. Hamzeh and R. Davoodi, *Cellulose*, 2015, **22**, 935-969.

5. Y. Habibi, L. A. Lucia and O. J. Rojas, *Chemical reviews*, 2010, **110**, 3479-3500.

6. Y. Habibi, *Chemical Society Reviews*, 2014, **43**, 1519-1542.

7. A. Salam, L. A. Lucia and H. Jameel, *Acs Sustainable Chemistry & Engineering*, 2013, **1**, 1584-1592.

8. V. Coccia, F. Cotana, G. Cavalaglio, M. Gelosia and A. Petrozzi, *Sustainability*, 2014, **6**, 5252-5264.

9. B. Sun, Q. Hou, Z. Liu and Y. Ni, *Cellulose*, 2015, **22**, 1135-1146.

10. M. Roman and W. T. Winter, *Biomacromolecules*, 2004, **5**, 1671-1677.

11. C. Campano, A. Balea, A. Blanco and C. Negro, *Cellulose*, 2016, **23**, 57-91.

12. S. H. Osong, S. Norgren and P. Engstrand, *Cellulose*, 2016, **23**, 93-123.

13. T. Taipale, M. Osterberg, A. Nykanen, J. Ruokolainen and J. Laine, *Cellulose*, 2010, **17**, 1005-1020.

14. I. Gonzalez, S. Boufi, M. A. Pelach, M. Alcala, F. Vilaseca and P. Mutje, *BioResources*, 2012, **7**, 5167-5180.

15. S. R. D. Petroudy, K. Syverud, G. Chinga-Carrasco, A. Ghasemain and H. Resalati, *Carbohydr. Polym.*, 2014, **99**, 311-318.

16. A. Balea, N. Merayo, E. Fuente, M. Delgado-Aguilar, P. Mutje, A. Blanco and C. Negro, *BioResources*, 2016, **11**,

3416-3431.

17. A. Balea, Á. Blanco, M. C. Monte, N. Merayo and C. Negro, *BioResources*, 2016, **11**, 8123-8138.

18. M. Delgado-Aguilar, I. Gonzalez, M. A. Pelach, E. De La Fuente, C. Negro and P. Mutje, *Cellulose*, 2015, **22**, 789-802.

19. A. Balea, M. C. Monte, C. Negro and A. Blanco, *Environmental Science and Pollution Research*, 2017, **24**, 5049-5059.

500246 Fabrication of Functionalized Aerogels from Cellulose and Whole Biomass for Absorbing Formaldehyde from Indoor Air

Abstract:

Aerogel is an extremely light material with porous network structure, low density and high specific surface area. These properties bring it many prospective applications such as gas storing medium, gas sorbent, and gas filter. In the present study, a novel method was developed to fabricate cellulose/biomass â€“ chitosan blended aerogel as functionalized biosorbent to remove formaldehyde from indoor air. The aerogel was prepared from cellulose/biomass and chitosan in concentrated LiBr solution via a process consisting of dissolution/dispersion, gelation, washing, solvent exchange, and drying. Because of the high porosity, large surface area, and free amino groups, the resultant aerogel can react instantaneously with formaldehyde and remove it irreversibly from indoor air. The results indicated that at cellulose â €“ chitosan ratio of 2:1 (w/w), the resultant blended aerogel had density of 25 mg/cm^3 and specific surface area of 245 m^2/g. In a stationary chamber with initial formaldehyde concentration of 25 mg/m^3, the functionalized aerogel adsorbed 86% and 97% of the formaldehyde within 1 and 4 h, respectively, with an adsorption capacity of approximately 53 mg formaldehyde/g aerogel.

497301 Cellulose-Based Injectable Hydrogel Composite for pH-Responsive Drug Delivery

Abstract:

Cellulose-based injectable hydrogel composite for pH-responsive drug delivery

Zhaohui Tong[*,1], Nusheng Chen[1], Ling Chen [2], and Wilfred Vermerris[3]

[1]Department of Agricultural & Biological Engineering, University of Florida, PO Box 110570, Gainesville, FL 32611-0570, United States

[2] Department of Pediatrics, University of Florida, PO BOX 100215, Gainesville, FL, 32610-0215, United States

[3]Department of Microbiology and Cell Science & UF Genetics Institute, PO Box 103610, Gainesville, FL 32611

Abstract: Localized drug delivery, and prolonged, stimulus-driven release are approaches to maximize the function of drugs and minimize their side effects. This study developed a novel cellulose-based injectable hydrogel composite system containing pH responsive poly (ethylene oxide)-block-poly (2-(diisopropylamino) ethyl methacrylate) (PEO-b-PDPA) copolymer micelles for localized delivery and pH-triggered release. First, covalently in situ generated hydrogels were constructed by mixing hydrazide-modified carboxymethyl cellulose (CMC-NH_2) with oxidized carboxymethyl cellulose (CMC-CHO). The resultant hydrogels could be well tuned in the aspects of gelation time, degradation profile, and mechanical properties. Then, the pH-responsible copolymer PEO-b-PDPA was introduced to enhance the loading of hydrophobic substances in hydrogels as well as a pH-triggered release, which was demonstrated by fluorescent spectroscopy using Nile Red as a model drug. In addition, the results demonstrated that the hydrogel itself could act as a second diffusion barrier to reduce the drug release rate. The cell viability tests showed that both CMC-NH2 and hydrogels possessed excellent biocompatibility, while CMC-CHO was cytotoxic at a relative high concentration.

499543 Biodegradable Seed Wraps for Sustained Release of Pesticides for Crop Protection in Sub Saharan Africa

Abstract:

Plant parasitic nematodes(PPNs) are one of the major constraints on yield and production of food crops throughout the world. While abamectin has broad nematicidal activity, its poor mobility in soils compromises its efficiency against nematode infestation. Using abamectin as a model pesticide, we present an innovative and cost effective approach for crop protection from PPNs, using a lignocellulosic matrix 'wrap and plant' methodology. In our 'wrap and plant' approach, we have used abamectin loaded matrices as wraps for seeds to be planted in the soil. To better understand the effect of various processing parameters and nature of the pulp on final properties of the matrix, we have produced lignocellulose matrices using pulp refined at various time intervals. Resulting hand sheets are characterized by the controlled release of abamectin which is incorporated into the matrices through sorption. Pulp refining time as well as various physical and chemical properties of the matrix are found to affect its tendency to let the germinating roots penetrate it and also the release rate of abamectin. We perceive that by varying refining time of the pulp, we can obtain matrices that facilitate in slow and sustained release of abamectin which results in long term nematicide protection to the growing plant roots. We have used these results to develop pilot scale paper rolls which are currently used in field trials in various regions of Africa and initial field trials demonstrate promising results regarding better crop protection in nematode infested soils in Benin and Kenya.

497553 Design, Fabrication, Testing, and Operation of a Continuous Reactor for Hydrothermal Carbonization

Abstract:

Hydrothermal carbonization (HTC, thermal hydrolysis, or wet torrefaction) is a treatment process which converts moist feedstocks into homogenized, carbon rich, and energy dense solid fuel, called hydrochar. One of the main advantages of HTC compared to other thermochemical treatment processes is the use of moisture as reaction medium and reactant, thereby precluding the need for drying prior to treatment. Thermodynamic properties of water change greatly in the subcritical region from 180-280 °C, and as a result, subcritical water behaves as a non-polar solvent and mild acid and base catalyst simultaneously. When subjected to HTC biomass undergoes rapid hydrolysis and other reactions, releasing oxygen-containing volatiles and producing a highly hydrophobic hydrochar.

Although HTC offers a relatively simple and straightforward solution to process diverse biomass feedstocks, the requirements of high pressure and high temperature make the process complex and potentially costly to design and operate. The lab-scale batch process has already been demonstrated in various laboratories around the world, but batch process is not cost-effective for industrial-scale deployment. A continuous process offers a relatively small footprint, energy recovery and efficiency, and notably, economies of scale.

In this study, a bench-scale continuous HTC reactor system has been designed, constructed, commissioned, and operated with cellulose and with whole biomass feedstocks. The maximum temperature and pressure of the reactor are 250 °C and 40 bar. The throughput of the reactor system was maintained at 0.3 L/min with solids loading of ~5%, while the reaction time was maintained at 5 min. The basic reactor configuration is a section of straight pipe, and numerous engineering challenges were encountered during design and commissioning. The primary challenges are related to instrumentation durability, temperature control, and flow control.

Results are presented for hydrothermal conversion of microcrystalline cellulose. Both solid and liquid products were tested for their physico-chemical properties and compared with the corresponding products from batch process generated in a 2L batch Parr reactor. It was found that temperature and pressure were stable during operation and products are relatively similar to that of batch process. Similar results were found for HTC of cow manure, when comparing solid yield, char fuel value and total organic content in the aqueous phase, when corrected for water volume.

495138 Fate of Nitrogen during Hydrothermal Treatment of Septage

Abstract:

Septic tanks are a popular method in rural areas to treat human waste. Waste enters the septic tank and settles so that solids collect in the bottom of the tank. Here organic matter is aerobically digested. Digestion only occurs partially, and is variable with each design of tank and the conditions of the tank (such as temperature and extent of cleaning). Septic tanks offer many advantages, such as the containment of the digestate and the containment of human waste. However, the resulting solid and liquid products must be pumped and disposed of, or further treated. One concern with the use of septic tank is the long term effects of untreated pollutants that are only present in the microgram/liter range, which have been shown to have serious impacts on aquatic ecology of surrounding water systems [1],[2]. Failure of septic tanks can also result in contamination of surrounding water sources with the various macronutrients present in the digestate and effluent of the tank.

Hydrothermal treatment has been investigated as a possible method for the treatment of municipal solid waste and sewage sludge [3]. Septage, or the combination of both septic sludge and effluent, can also be treated by hydrothermal means. Hydrothermal treat is an attractive option for additional septage treatment due to its high water content; there is no need for any drying or separation for solid treatment. More so, the high nitrogen content of septage can be recovered and reused as fertilizer. In this study septage was collected from local health department and hydrothermally treated. A full nitrogen balance was performed between soluble nitrogen species, as well as solid nitrogen content determined by CHNS. Experimental trials for this treatment used a batch reactor with a reaction time of 30 minutes. It was found that hydrothermal treatment at temperatures between 180-260 °C results in increased nitrogen in the process liquid compared to the untreated septage. In addition, treatment results in increased ammonia concentration in the liquid phase and a pH between 6 and 8. This generation of soluble nitrogen species is a result of the solid phase decomposing as it is heated.

References

[1]. Schwarzenbach, René P., et al. "The challenge of micropollutants in aquatic systems." *Science* 313.5790 (2006): 1072-1077.

[2]. Withers, Paul JA, et al. "Do septic tank systems pose a hidden threat to water quality?." *Frontiers in Ecology and the Environment* 12.2 (2014): 123-130.

[3]. Zhao, Peitao, et al. "Clean solid biofuel production from high moisture content waste biomass employing hydrothermal treatment." *Applied Energy* 131 (2014): 345-367.

491306 Hydrothermal Treatment of Paper Mill Sludge

Abstract:

Each year the pulp and paper industry generates large amounts of effluent treatment sludge. The sludge is made up of three different fractions namely, primary, secondary, and tertiary sludge. Depending on the layout of the sludge handling in the individual mill, these three sludges can either be extracted separately or mixed into a stream called mixed sludge. Usually secondary and tertiary sludge are mixed together. The primary sludge, often rich in fiber material and easy to dewater, can easily be used as fuel source. However, water is difficult to remove from mixed sludge.[1-2] Moreover, a low fuel heating value of mixed sludge results against economical combustion. Therefore, a new technology that converts mixed sludge to viable solid fuel may be required for mixed sludge valorization.

Hydrothermal carbonization (HTC) has received considerable attention during the last several years as a prominent alternative for upgrading biomass feedstock ranging from wood and algae to agricultural and forest residues and industrial or municipal wastes. One of the main advantages of HTC is the use of aqueous solutions under high temperature (180-260 °C) and autogenously generated pressure (5-50 bar).[3] As a result, moist and wet feedstocks are favorable for HTC.[4]

This goal of this study was to determine the effects of HTC conditions on paper mill sludge dewaterbility and fuel value. A 100 ml Parr batch reactor was be used for HTC. Reaction temperature was varied from 180-260 °C, while reaction time was varied from 30 min-1h. The solid hydrochars were analyzed for energy content, ultimate analysis, FTIR, and NMR. Meanwhile, the process liquid is characterized with GCMS, HPLC, UV-VIS, NMR, and pH. The energy content of the hydrochar was increased with reaction temperature and longer reaction time as a result of the decrease in oxygen content. The resulted hydrochar is more hydrophobic than the feedstock. Therefore, dewaterbility is enhanced with the hydrothermal treatment of paper mill sludge.

References

1. Makela, M.; Benavente, V.; Fullana, A., Hydrothermal carbonization of industrial mixed sludge from a pulp and paper mill. *Bioresource Technol* **2016,** *200*, 444-450.

2. Makela, M.; Yoshikawa, K., Simulating hydrothermal treatment of sludge within a pulp and paper mill. *Appl Energ* **2016,** *173*, 177-183.

3. Reza, M. T.; Andert, J.; Wirth, B.; Busch, D.; Pielert, J.; Lynam, J. G.; Mumme, J., Hydrothermal Carbonization of Biomass for Energy and Crop Production. *Applied Bioenergy* **2014,** *1* (1).

4. Reza, M. T.; Coronella, C.; Holtman, K. M.; Franqui-Villanueva, D.; Poulson, S. R., Hydrothermal Carbonization of Autoclaved Municipal Solid Waste Pulp and Anaerobically Treated Pulp Digestate. *Acs Sustain Chem Eng* **2016,** *4* (7), 3649-3658.

486662 Structural Analysis of Humins Formed in the BrÃ¸nsted-Catalyzed Dehydration of Fructose

Abstract:

Humins are carbonaceous and polymeric by-products formed during acid-catalyzed hydrothermal processing of sugars to 5-hydroxymethylfurfural (HMF). These are formed via cross-polymerization of HMF with sugars in acidic solutions, and cause significant carbon loss and unfavorable process economics of biorefineries. The deposition of these solid particles on the reactors and processing equipment also increases the maintenance costs. Currently humins are combusted to supply heat in biorefineries. Minimization of humins formation or their valorization is essential for improving the process economics of the biorefineries. In this work, we developed methods to understand microstructure of humins through solubility experiments. First, we have conducted dissolution experiments in various solvents and correlated the solubility data to molecular properties of solvents to develop suitable descriptors. By doing this, we have identified solvents for high solubility of humins at room temperature and provided insights into the mechanism of humins dissolution. Next, Liquid Chromatography- Mass Spectrometry (LC-MS) was used to determine and compare the masses of components in the humins solubilized in different solvents. Finally, a six stage dissolution experiment was done to investigate the structural inhomogeneity of humins. Both Fourier Transform Infrared Spectroscopy (FT-IR) and LC-MS were used to characterize the solubilized and insoluble humins obtained at each stage. Based on our results, an inhomogeneous structure where macromolecular and molecular components are connected through weak forces, was proposed.

Catalysis Center for Energy Innovation

Super Oleophilic and Water Repellent Graphene Oxide Sponge for Catalytic Coupling of Furans and Oil Absorption

Saikat Dutta

Advisers: Basu Saha, Dion G. Vlachos

Catalysis Center for Energy Innovation

University of Delaware

Newark, Delaware

Session: Hydrothermal Carbonization

November 2, 2017

AIChE Annual Meeting 2017, Minneapolis, MN

U.S. DEPARTMENT OF ENERGY | Office of Science

CCEI is an Energy Frontier Research Center funded by the U.S. Department of Energy, Office of Basic Energy Sciences

Catalysis Center for Energy Innovation

Why Graphene Sponge

- Graphene is generally hydrophobic, can be tuned to hydrophilic

Hydrophobic properties

Graphene@SiC

CA 92°

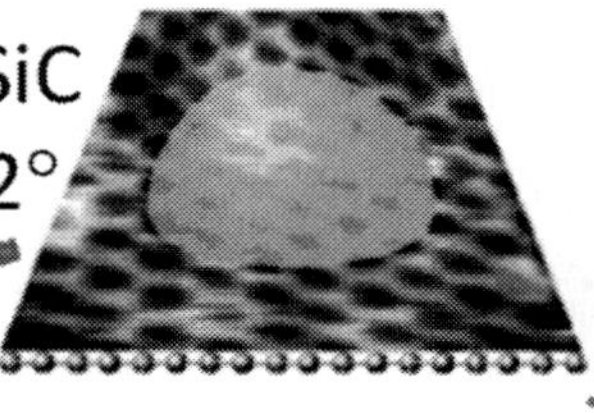

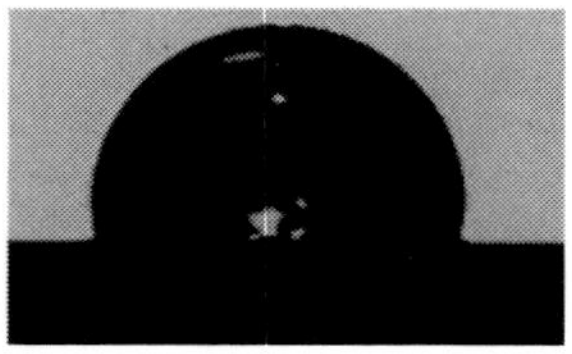

SiC CA 69°

O_2 plasma

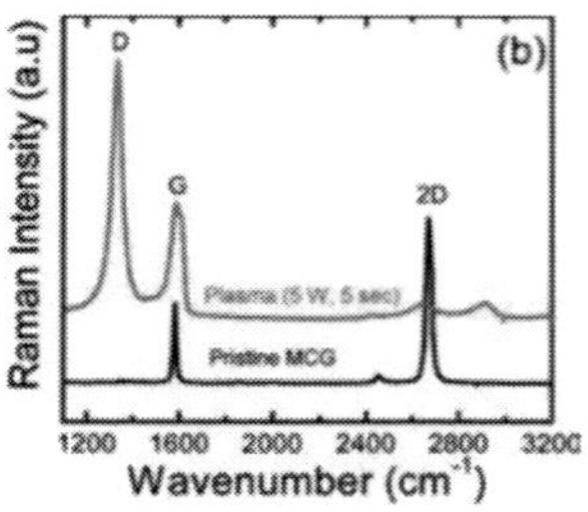

- Thickness Independent Hydrophobicity

Energy Environ. Sci. **2012**, *5*, 7908

Langmuir **2010,** *26*, 3798

- More defects, more hydrophilic
- GO sponges are not explored extensively

2

Sponge Materials: Why Is It Important

Marine Oil Spill

Threats to the planet

- High frequency
- Long-lasting adverse environmental impacts
- Significant economic impact

Current Methods

- Skimming on calm water
- Chemical dispersants spread

Sponge Materials

carbon-based aerogels

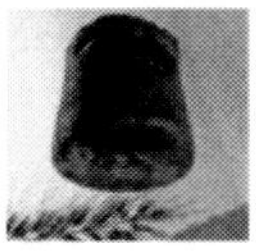

Adv. Mater. **2013**, *25*, 2554

porous boron-nitride

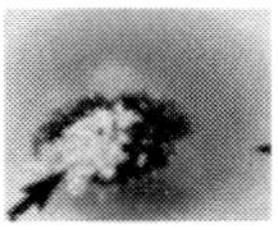

Nat. Commun. **2013**, *4*, 1777

modified commercial sponges

Energy Environ. Sci. **2012**, *5*, 7908

GO-wrapped sponge

Nat. Nanotechnol. **2017**, *12*, 434

3

Graphene Oxide Sponge (GOS)

- Explore hydrophobic and hydrophilic behaviors of GOS

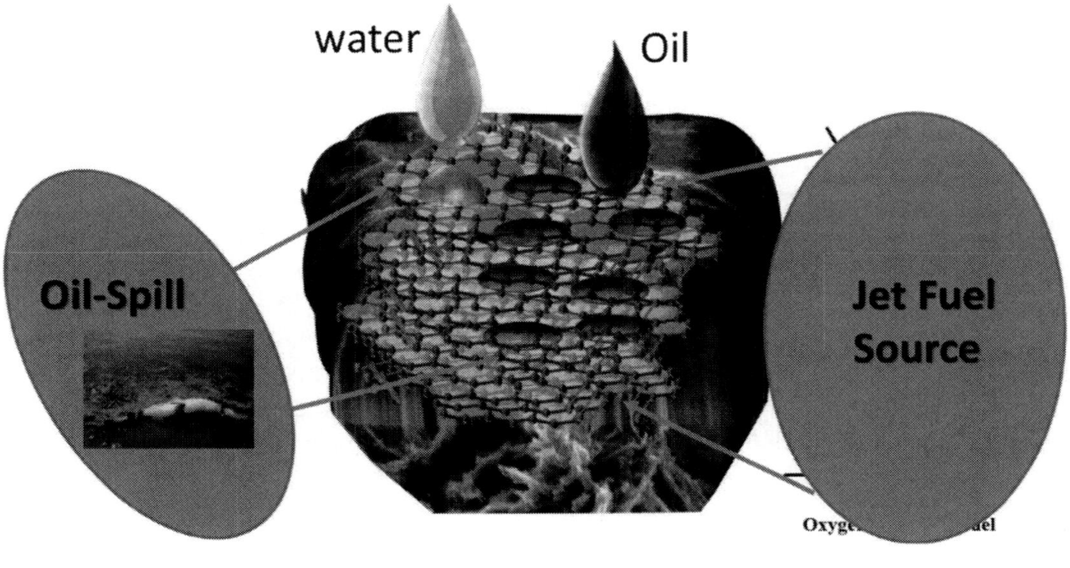

4

Hydrophilic IGO to GOS

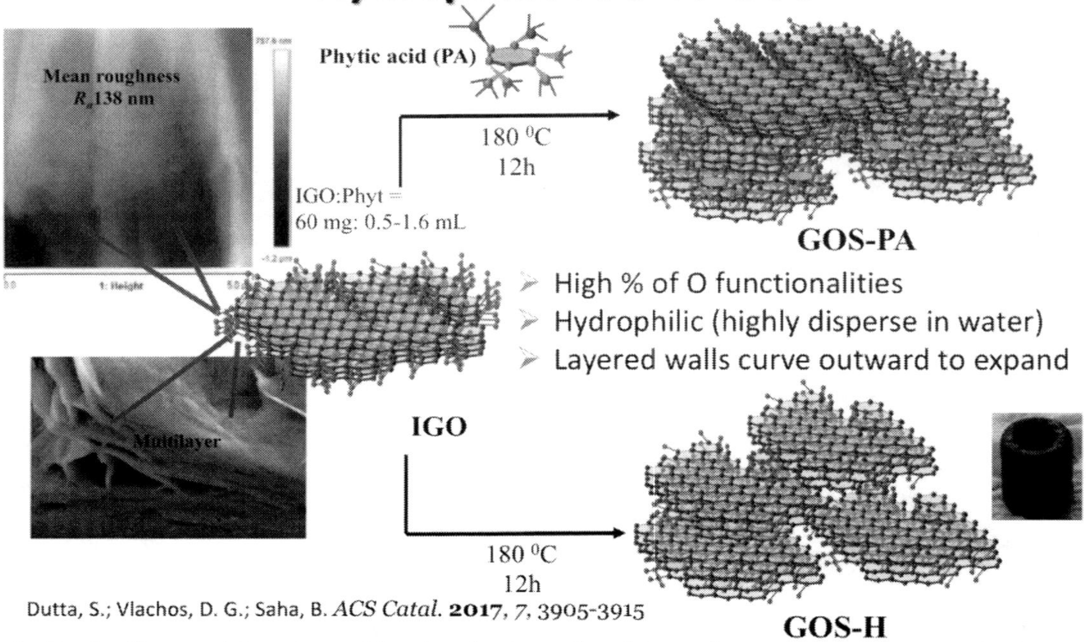

- High % of O functionalities
- Hydrophilic (highly disperse in water)
- Layered walls curve outward to expand

Dutta, S.; Vlachos, D. G.; Saha, B. *ACS Catal.* **2017**, *7*, 3905-3915

- Hydrothermal treatment is simplest method to reduce surface oxygen (gain hydrophobicity) compared to cross-linking/freeze-drying

5

Hydrophobicity of GOS from XPS C1s Peak

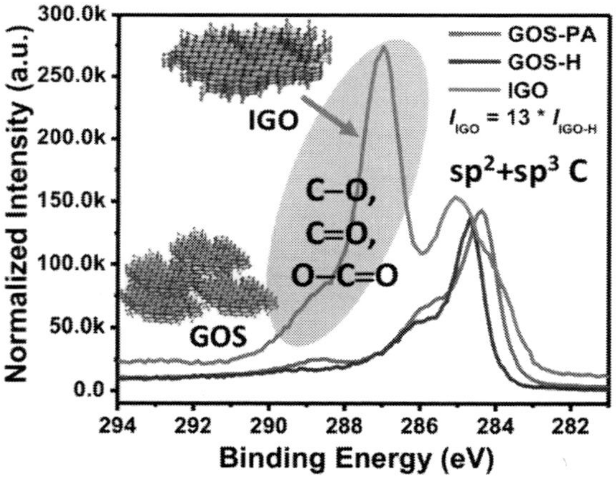

More surface oxygen (hydrophilic)

Less surface oxygen (hydrophobic)

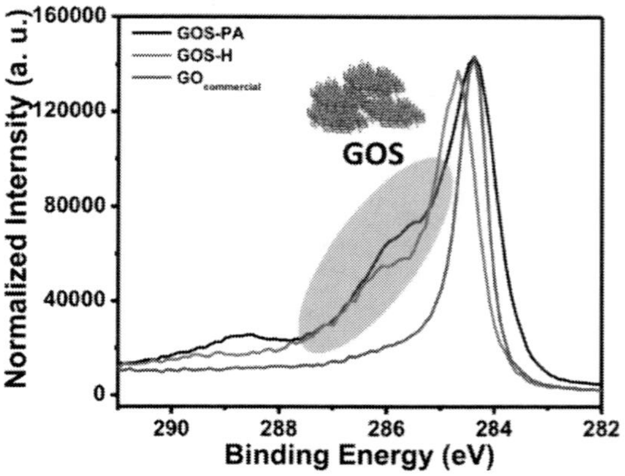

- Increase in hydrophobicity: selective for oil/organic absorption

- Surface oxygen GOS > GO$_{commercial}$
- GOS more reactive

6

Defect Analysis: Raman Spectra

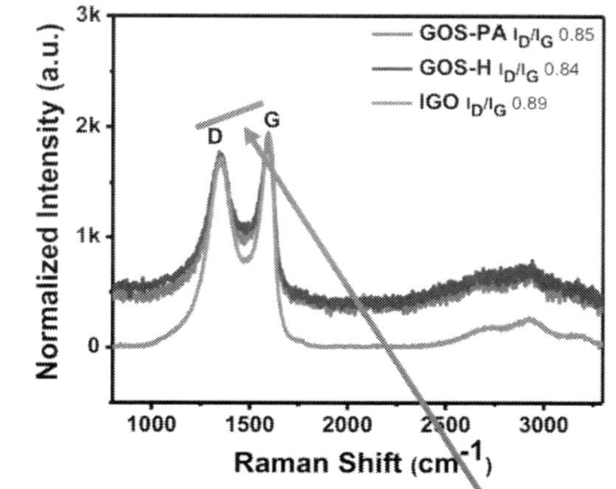

O_2 plasma (introduction of defect)

G@SiC

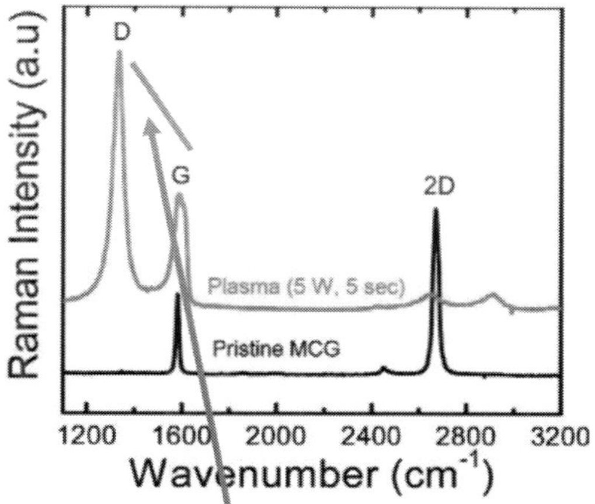

Langmuir **2010**, *26*, 3798

- GOS and IGO comparable crystalline defect
- G contains less crystalline defect before plasma treatment

Raman Imaging of D and G Bands

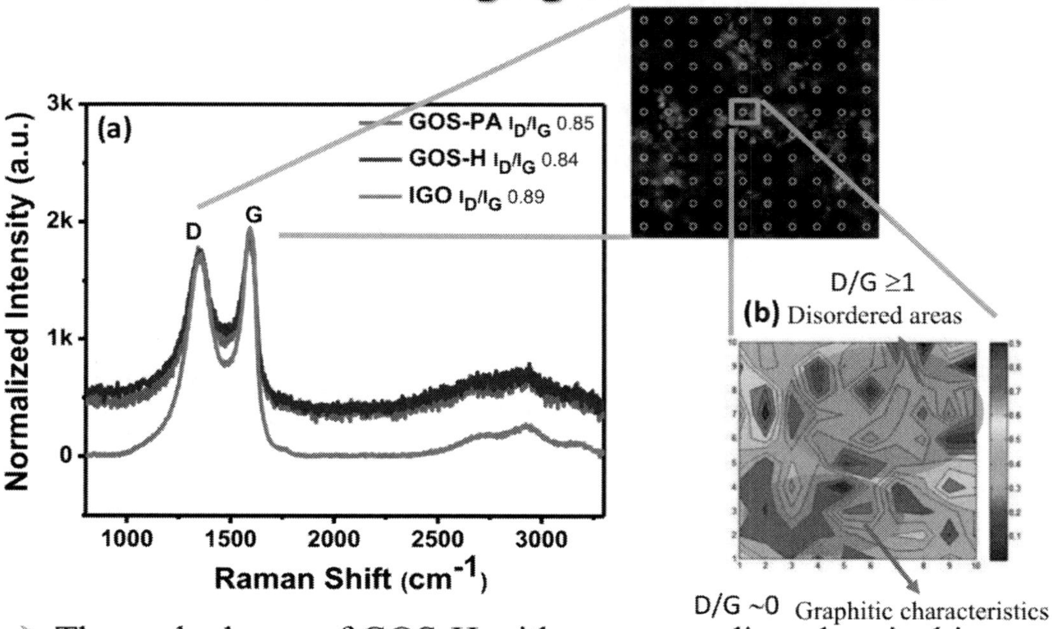

- The probed area of GOS-H with corresponding chemical image displaying the ratio of D-band and G-band intensity (b)

SEM Characterization of GOS-PA and GOS-H

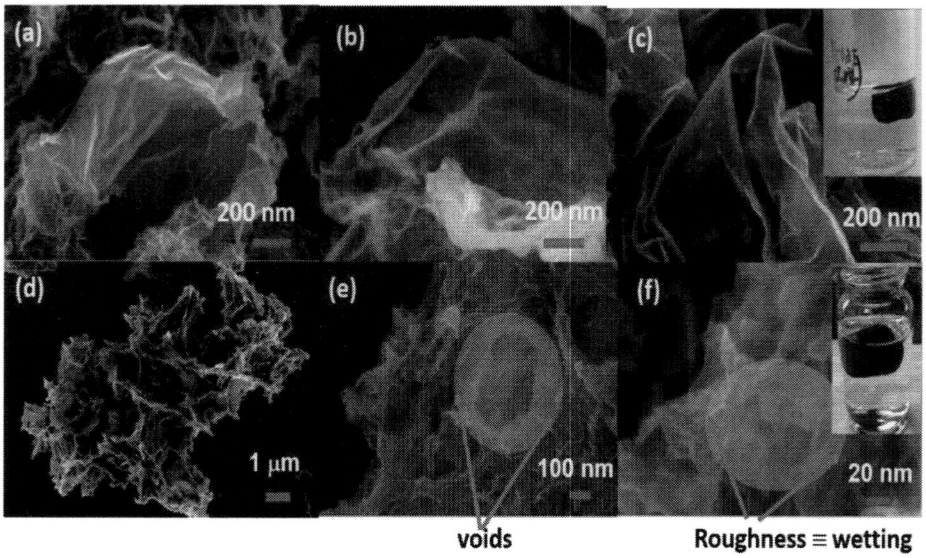

➢ GOS-H contains more rough surface as compared to GOS-PA

9

Water-Contact Angle of GOS

Sample	Contact Angle (Degrees)	Error (%)
HOPG (std)	67.3	3.3%
GOS-H	86.5	0.7%
GOS-PA	84.2	1.8%

➢ GOS hydrophobic with contact angle ~90°
➢ Enhancement of contact angle is countered by surface oxygen

10

Oil Absorption by GOS-H

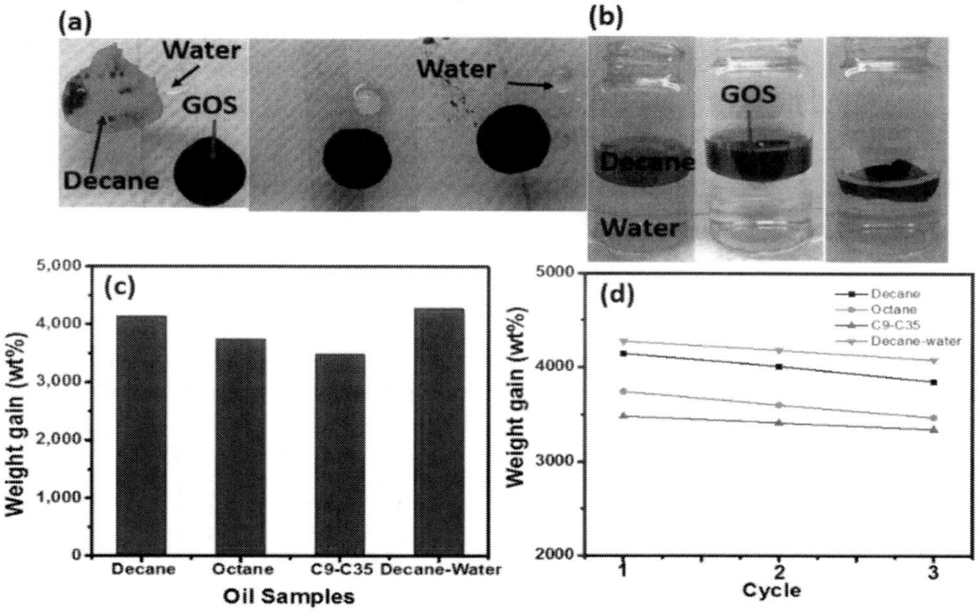

- Selective oil absorption by GOS in 15 min
- Exhibit recyclability for oil absorption

Origin of Hydroxyalkylation-Alkylation (HAA) Catalysis

- Surface-oxygen active sites of IGO enables tailor the C-numbers of lignocellulosic fuel precursors

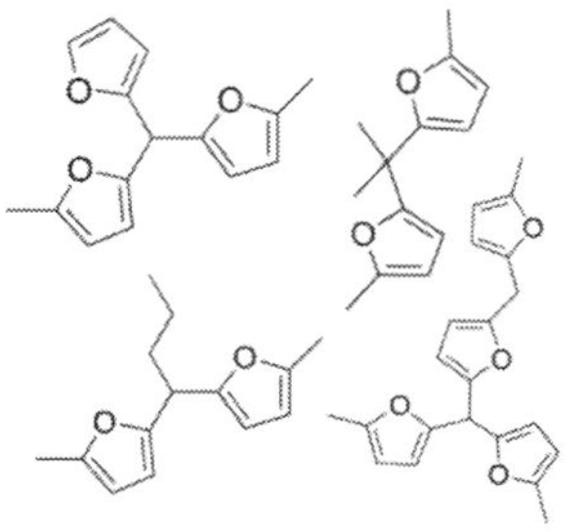

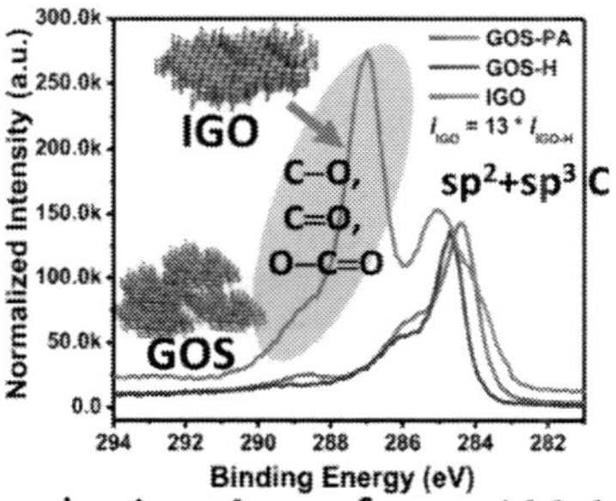

Entry	C (%)	O(%)	P(%)	S(%)
IGO-H	73.4	25.1		1.4
IGO-PA	72.5	24.2	2.1	1.2
IGO	59.2	36.6		3.1
HOPG	98.2	0.41		

- C13-C21 furan-derivatives from HAA with IGO

Dutta, S.; Saha, B. *ACS Catal.* **2017**, *7*, 5491-5499

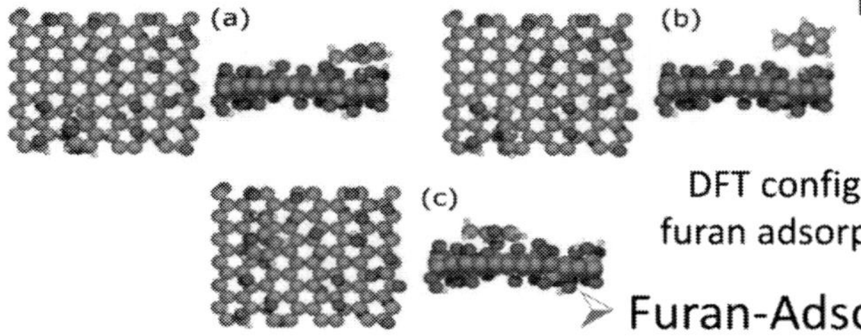

DFT configuration of furan adsorption on IGO

- Furan-Adsorption on a Oxygenated Surface

Dutta, S.; Vlachos, D. G.; Saha, B. *ACS Catal.* **2017**, *7*, 3905-3915

HydroxyAlkylation-Alkylation (HAA) of Furans

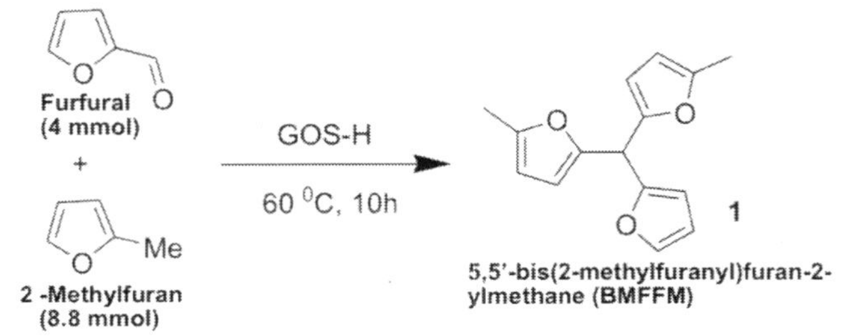

GOS-H recycle

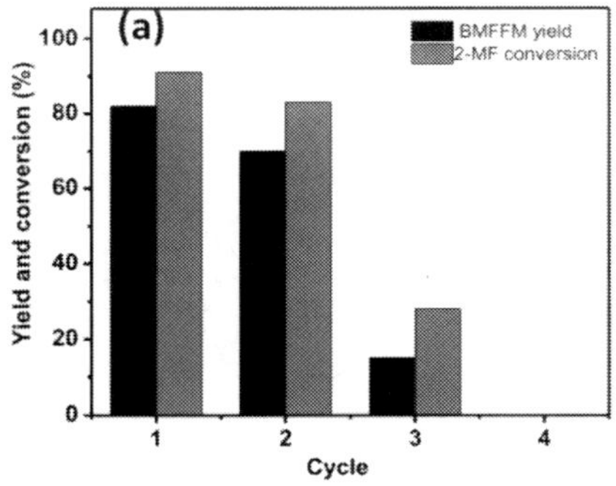

- Carbon extension of furans to desired fuel precursors
- Intermediate surface oxygen % with hydrophilic behavior GOS-H for HAA
- Loss of activity due to further reduction of surface oxygen (XPS)

13

Summary

- GOS derived from aqueous dispersion of IGO.
- GOS contains surface roughness responsible for wettability
- Raman imaging, XPS analysis, and SEM were used for characterization
- Percentage of surface oxygen in GOS is higher than graphene or rGO sponge
- GOS-H is effective for HAA reactions for furans
- GOS-H offers selective oil absorption from oil-water mixture

Moving Forward

- Technique for cross-linked GOS for mechanical stability

14

Catalysis Center for Energy Innovation

Acknowledgements

Joseph P. Smith (Raman Spectroscopy)
Prof. Karl S. Booksh (Raman Spectroscopy)
Dr. Weiqing Zheng (SEM,TEM imaging)
Rachel Pupillo (XPS)
Dion G. Vlachos (insightful discussions)
Jeff Everhart (GC analysis)

U.S. Department of Energy, Office of Science, Office of Basic Energy Sciences, under Award No. DE-SC0001004

U.S. DEPARTMENT OF ENERGY | Office of Science

15

502391 Simultaneous Isomerization and Reactive Extraction Followed By Back Extraction of Sugars from Biomass Hydrolysate for High Purity and Yield of Ketose Sugars

Abstract:

Biomass hydrolysates are often low in sugar concentration and contain contaminants associated with the pretreatment method and sugar and lignin degradation. Cost-effective and high-yield downstream processing requires sugar streams at higher concentration, free of contaminants generated during the pretreatment process. Instead of removing contaminants from the hydrolysate, we have taken the approach of removing sugars from the hydrolysate, leaving contaminants behind. In the process of removing sugars, we enzymatically isomerize them from their aldose (glucose, xylose) to their ketose (fructose, xylulose) forms. These ketose sugars are ideally suited for production of furans â€" HMF from fructose and furfural from xylulose.

Unfortunately, the isomerization reactions do not have favorable equilibria with respect to ketose formation. To overcome this limitation, we have previously developed a method of simultaneous-isomerization-and-reactive-extraction (SIRE) to produce the ketose isomer in high yield. We have demonstrated that with a xylose/xylulose mixture, SIRE followed by 2-stage back-extraction (BE) allows high yield recovery of xylulose in nearly pure form.

For mixtures of xylose and glucose found in biomass hydrolysate, separation and purification of the ketose sugars (xylulose and fructose) from the hydrolysate will be more complex. The extraction affinities of glucose and fructose are lower than xylulose, but higher than xylose. SIRE-BE has not been previously evaluated for biomass hydrolysates with high sugar concentrations (up to 90 g/L) for recovery of a high purity, mixed C5 and C6 ketose product stream. We have developed strategies for high purity recovery of both xylulose and fructose from mixed sugar streams using SIRE-BE. We will present these strategies as well as xylulose and fructose yields and purity obtained under different experimental configurations.

AUTHOR INDEX

AUTHOR INDEX